NASA SP-4106

Aiming at Targets

The Autobiography
of
Robert C. Seamans, Jr.

Robert C. Seamans, Jr.

The NASA History Series

NASA History Office
Office of Policy and Plans
NASA Headquarters
Washington, DC
1996

Library of Congress Cataloging-in-Publication Data

Seamans, Robert C.
    Aiming at Targets: The Autobiography of Robert C. Seamans, Jr./
Robert C. Seamans, Jr.
    p. cm.—(The NASA History Series) (NASA SP: 4106)
    Includes index.
    1. Seamans, Robert C. 2. Astronautics—United States—Biography.
3. United States. National Aeronautics and Space Administration—
Officials and employees—Biography. 4. United States. Air Force—
Officials and employees—Biography. I. Title. II. Series.
III. Series: NASA SP: 4106.
TL789.85.S43A3 1996
629.4'092—dc20
[B]

CONTENTS

i

ISBN 0-16-048907-5

9 780160 489075

90000

FOREWORD

Willis H. Shapley

B ob Seamans originally was inspired to write this book for his family
and friends. That is a large audience. By his own count his immediate family numbers twenty-four, not counting brothers and cousins
and their families. His friends are uncounted but surely run to hundreds. As one of them and as a colleague at NASA, I am pleased and
honored that he asked me to write this foreword.

While written in Bob's unique and informal style, this autobiography has significance for many readers beyond his large circles of family and friends. Leaders and students of large, complex technological
endeavors should be able to learn much from reading how Bob faced
the daunting technical and management challenges in his career. As the
title of this book implies, Bob has always set high goals for himself and
then kept his eyes focused on both the necessary details and the
broader picture. His ability to shift smoothly among jobs that required
seemingly disparate abilities and skills speaks volumes about his
insight, dedication, and enthusiasm for achievement.

The book spans a truly remarkable life story. Bob first takes us
through his growing up, education, and early professional and family
life. Next he focuses on the crucial years when he was the general manager of NASA. Then he moves on to his career in the top jobs at the
Air Force, the National Academy of Engineering, and the Energy
Research and Development Administration. Finally, he touches on his
later leadership activities in the academic and business worlds.

Aiming at Targets is a series of fascinating topical vignettes cover-
ing his professional life. Taken together, like broad brushstrokes in an
impressionist painting, they give a better picture of Bob Seamans and
his work than a detailed recitation of facts and dates could hope to do.

This is a cheerful account of an interesting and successful career. The book is full of good stories, with many memorable characters. Like the proverbial sundial, it counts the sunny hours. It is a good read.

But it has its serious side. Bob's career wasn't all fun. The Apollo 204 fire, which killed three astronauts, was a terrible climax to his time at NASA. As one who lived through those days with him, I can recall the trauma and special sense of responsibility he felt. His account of this period and of the sad deterioration of his relationship with his boss, Jim Webb, is both fair and generous. Those were not happy times, but they should not be allowed to overshadow the fact that in his seven years at NASA, Bob Seamans led the agency to its first successes and laid the groundwork for the greater successes that came later.

Also on the serious side, while secretary of the Air Force, Bob had to face policy differences on the Vietnam War, both on the job and within his family. He writes of this frankly and kindly. As the book moves on through Bob's career, there are explanations, spiced with lively anecdotes, of what he did in each of his jobs. These are well amplified in the appendices. All of this is written with unfailing modesty, which understates Bob's accomplishments and makes it all look easier than it was.

Most of all, what comes through is the happy and productive life of a fortunate man—fortunate in his abilities, in opportunities to apply them, and in his wonderful wife and family. They, as well as everyone interested in the work with which Bob Seamans was involved, are fortunate to have this engrossing personal account.

Preface and Acknowledgments

*F*rom time to time, friends and family have suggested that I write a book about my professional life. Until recently I have resisted the pressure, although I have participated in a variety of efforts to help document my experience while serving in government.

These historical efforts were mostly oral interviews conducted at key points in time. Walter Sohier, Deputy General Counsel of NASA, interviewed me along with others in our agency soon after the death of President Kennedy. After I retired from NASA, NASA historian Eugene Emme and I had an exchange. Similar exit interviews took place upon my leaving as secretary of the Air Force and as administrator of the Energy Research and Development Administration. In all cases, a transcript was delivered to me for editing. These transcripts are now in the archives of the Massachusetts Institute of Technology (MIT).

For interested researchers, the archival collection in the NASA Headquarters History Office has numerous papers, memos, photographs, and other records related to my tenure there. Yale University, MIT, and the Air Force Historical Research Center at Maxwell Air Base in Alabama also house some key papers from my career.

I served at NASA for over seven years and four months, working first in the Eisenhower administration for Administrator T. Keith Glennan and then in the Kennedy and Johnson administrations for James E. Webb. After we all had retired from NASA, Jim, an ardent historian, sold Keith and me on a program for interviewing thirty or so key individuals at NASA. This record, obtained in interviews with historian Martin Collins, is being edited for future historians by the Smithsonian Institution's National Air and Space Museum. These edited transcripts (in my case resulting from twenty-nine hours of questioning) may be obtained from the archivist at the museum.

Retirement, I've found, is not a simple process. One doesn't simply walk through a gateway into a beautiful garden and discover that one now has time to meditate. But several years ago, I realized that I did have more control over my activities. This, coupled with the ongoing pressure to produce a book and the archival experience just described, has led me inexorably to this autobiography.

I realized that I had two fundamental reasons for writing, personal and professional. First, I wanted my family and friends to have a readable document that would tell them in reasonable depth about my remarkable ringside view of the space program. Second, I wanted to rough out for professionals my firsthand view of the key decisions leading to humanity's landing on the Moon. Many books have been written and TV documentaries produced about this subject, and I have no quarrel with most of these. However, like the blind men describing the elephant, most have told only part of the story. How was the program perceived by those who managed it, and how did they execute their responsibilities? What is known about the forces that motivated four U.S. presidents and leading members of Congress to embark on such a major scientific and technical enterprise? I have attempted to answer these larger questions from the perspective of my place as a participant.

Some felt that my autobiography should be stripped of family anecdotes and written for a larger audience. I chose to include these stories because my family life has always been of the greatest importance to me. Not only has it been interwoven with my professional life, it has fed and nurtured that life. I could not write the story of "my life" without giving my family a prominent place.

Since my first years at MIT, my wife, Gene, has played a huge role—as supportive friend and excellent counselor—in all of my endeavors, including the writing of this autobiography. She has become familiar with all of the enterprises with which I've been involved. Most important of all, she has been the fairy godmother holding our family together. Her "Seamans Messenger" is eagerly awaited not only by members of our immediate family, but also by brothers, sisters, nieces, and nephews. She has been a motivator and inspiration for all of us.

Our five children also have been key actors in this story. Each offered his or her suggestions and wise counsel as I wrote this manuscript. They and their spouses are now our closest friends. The twelve grandchildren for whom they are responsible are becoming fascinating individuals.

Outside the family, I want to thank Martin Collins, whose interviewing and dogged pursuit of the facts provided much of the basic material for the central NASA chapter of this memoir. Roger D. Launius, NASA's present historian, reviewed the manuscript when it was embryonic and encouraged me to continue. Roger, Louise Alstork, Nadine Andreassen, Stephen Garber, and Lee Saegesser in the NASA History Office also went to considerable effort, supplying factual information, photographs, graphic material, and editorial assistance. Webster Bull of Memoirs Unlimited was instrumental in editing and formatting an earlier version of this book. I thank Donna Martinez for interpreting my first draft of chapter three, "The Air Force Years," as well as W.S. Moody and the other volunteers in the U.S. Air Force Historical Support Office for providing important details. My sincere thanks also go to Jim Harrison and Julie Merrill at MIT's Draper Laboratory for assistance in preparing the map of Southeast Asia and to Richard P. Hallion and Herman R. Wolk at the Air Force History Office for their expert advice. I am also very grateful for the tireless efforts of the NASA Headquarters printing and graphics team, including Michael Crnkovic, Jonathan Friedman, Jim Harlow, O'Neil Hamilton, Craig Larsen, Jane E. Penn, and Kelly Rindfusz.

Finally, I recognize that there were many in government, industry, and academe who made this story come true. I have mentioned some in the text, but let me offer my thanks to three others who are not mentioned. Mary Traviss was my secretary for much of my time at NASA. Each day she told me where to go and what to do. She kept a detailed log of my activities and was able to convert my hieroglyphics into readable prose. David Williamson, my assistant at NASA, could think and write analytically. He could differentiate between a panic and a true crisis. He served NASA well and faithfully for many years after I left, in the face of serious medical problems. Finally, James Elms was an important player at NASA, serving as deputy to Robert R. Gilruth at the Johnson Space Center and director of the Electronic Research Center in Cambridge, Massachusetts. Much could be written about his important role in both capacities. But Jim was more than a line manager. He was a confidant who served me as a sounding board and advisor during critical times.

Willis Shapley is also mentioned in the text. In the early 1960s, he was the chief examiner of NASA in the Bureau of the Budget (now the

Office of Management and Budget). He was shrewd and thoughtful in his budget and management critiques. In 1965, he left the Bureau of the Budget, joined NASA, and became an important member of the NASA administrator's immediate team. When I recognized the need for a critical review of this memoir, I could think of no one more suitable. He offered helpful advice and wrote a perceptive, though perhaps too generous, foreword. I am truly grateful.

Life is an arrow—therefore you must know
What mark to aim at, and how to use the bow
Then draw it to the head, and let it go!
>   —*Henry Van Dyke (1852–1933)*
>   *poet and clergyman*

Aiming at Targets

CHAPTER ONE

Liftoff

H̲EROES, to me, are people who have tried to beat the odds, people who have made the most of what they have. In hockey, an example from my youth was the Boston Bruins' Eddie Shore. If the Bruins were down by a goal in the middle of the third period, the crowd would start chanting, "We want Shore! We want Shore!" Sooner or later, Eddie Shore would wind up from behind his own net, carry the puck the length of the ice, and (more often than not, it seemed) score the tying goal. Terrific!

In my chosen field of aeronautics, the Wright Brothers were early heroes, taking on the Smithsonian Institution. These were two men who made *bicycles*, but they went up against the Washington intelligentsia. Charles A. Lindbergh had a similar story, a nobody who succeeded despite great odds. Competing against several more experienced airmen, such as Navy flier Richard E. Byrd, World War I ace René Fonck, and stunt flier Clarence E. Chamberlin, in 1927 Lindbergh became the first pilot to fly nonstop from New York to Paris. Motivated in part by a $25,000 prize originally offered in 1919 by a New York hotelier from France, the twenty-five-year-old Lindbergh accomplished his amazing feat by insisting on exacting standards for his specially built airplane.

During World War II, tremendous heroes emerged as flyers. After the war, the great challenge was to break the sound barrier. Through my work at the Massachusetts Institute of Technology (MIT), I became involved in studying why an airplane became uncontrollable as it approached the speed of sound. What a thrill it was to listen to the tape recording of Chuck Yeager's commentary as he broke the sound barrier in October 1947! Then along came the U.S. space program, which took American heroism in flight to a whole new level. I was fortunate to have an inside view of the entire Apollo effort, an accomplishment that I believe will go down as one the most significant ever.

While I was secretary of the Air Force during the Nixon administration, my military assistant was William Y. Smith, an Air Force colonel from Arkansas. He came into my office every morning to brief me. After a while I noticed that every time he sat down, he kept his right leg extended in front of him. When I looked into his record, I saw that he had been shot down during the Korean War, had bailed out of his airplane, and had lost his leg, now replaced by a prosthesis. Further review of his record showed me that he was a West Pointer with a doctorate from Harvard University. By the time he retired, he had been promoted to four-star general. Willie Smith was a person who made the most of what he had, a true hero.

In our family we had an example of heroism ready at hand. General George S. Patton, Jr., married my wife Gene's aunt, Beatrice Ayer. World War II made General Patton a national hero. After the war, he came home for about a week, landing at Bedford Airport. Proud citizens lined virtually his entire route from Bedford to the Hatch Shell on the Charles River Esplanade in Boston, where he spoke that afternoon. Gene's parents, Mr. and Mrs. Keith Merrill, hosted a party for him at their home, Avalon, in Beverly Farms, Massachusetts, in the days following. Because meat was still unobtainable, we slaughtered some live chickens, due to war rationing. The next morning, a Sunday, the general spoke at a service at St. John's Church in Beverly Farms. Afterwards we had dinner at Woodstock, Gene's aunt's home in Prides Crossing. In the middle of dinner the general pointed his finger at me and said, "I want you to know that when I came into this family, my father-in-law was very much against war. I told him what I was trying to do in World War I, and he finally said, 'Well, just make sure that, if you're going to be a soldier, you be the best soldier you can be.' The most wonderful thing about this family," the general went on, "is that it energizes people to do the best they can."

I've certainly found that to be true. I couldn't possibly have married a more supportive person than Gene, and I am very grateful as well for the support of her mother, her brother Keith, her sister Romey, and "the old gentleman," her father. He could seem pretty tough, but he did wonderfully nice things for people, especially behind the scenes. When Gene and I were first married, we were living in a $65-a-month apartment and working pretty hard. He wrote me a note saying that he was proud of how well we were doing and that he thought Gene and I ought to have a chance to get out once in a while and do something by ourselves for fun. So he deposited $100 in our name at the Ritz, told us to go have a good time, and asked me to let him know when we needed more.

I have been lucky in love—and in work. Timing is very important. It just so happened that my professional capabilities meshed well with the timing of professional opportunities. Napoleon, when a soldier was brought to his attention for possible promotion, used to ask, "*Est il heureux?*" ("Is he lucky?") What he was looking for were men who, somehow or other, achieved their objectives. In that sense of the word, too, I have been pretty lucky. When there was a job to be done, I did not like to sit around debating; I liked to move ahead.

When I was eleven and going to the Tower School, our class was charged with selling advertisements for the school magazine, the *Turret*. I trotted around to the retail businesses in town and asked them all to advertise. The other kids selling ads found that wherever they knocked I had been there before them. I came into school with a whole sheaf of orders and by far the most change in my pocket. A full-page ad cost eight dollars.

I don't know why Charles Stark ("Doc") Draper picked me for the first of a series of projects at MIT during and after World War II. Perhaps I was lucky in that, but I usually did get the job done. Each project led to another with greater responsibilities. When the brass came up from the Pentagon to look at something we were developing, Doc used to say, "We're like little boys on the sidewalk watching the fire engines go by." What he meant was that something beyond us was happening and we were little more than observers, happy to be there. Bigger and bigger engines came past me. Finally there was one called NASA. In this case, I wasn't watching from the sidewalk. I was aboard and in the cab. But still like the little kid on the sidewalk, there were times when all I could do was watch with amazement.

Grandfather Bosson—An Early Family Influence

My family has always colored my outlook on life, and a variety of family members have served as inspirations for me throughout my career. I barely remember my mother's father, Albert Davis Bosson (1855–1926), but I would say that he was one of the first heroes in my life, once I became aware of his various accomplishments. He had his finger in a lot of different things—the Hood Rubber Company, the Naumkeag Mills in Salem, the Boston and Lockport Block Company (which made pulleys, originally for sailing vessels and more recently for cargo ships and oil rigs). He was a founder of the County Savings Bank, a relatively small institution in Chelsea, Massachusetts. He also served as a judge on the local court. Apparently, people didn't worry as much about conflicts of interest in those days.

My grandfather died when I was seven years old, having suffered from a bad heart for many years. He used to spend summers in Europe taking hot baths and other "cures." My parents told me that, upon his annual return from Europe, he always seemed worse than when he had left to go overseas. He would return to his apartment and go to bed. After three or four days of this, he would start picking up the phone and calling business associates. Then he might have a board meeting or two in his bedroom, and before long, he would be back in his chauffeured car finding out what was going on. With this involvement, he would come alive again. To me, this has always been an interesting commentary on the importance of remaining active.

I wasn't aware of all of my grandfather's business dealings, but I know that he played the piano, and I do remember that he taught me the Lord's Prayer. He was always reading four or five different books—a novel, a book of poetry, one on history, a great variety of things. He liked to work his way through all of them simultaneously.

My Parents

My mother's maiden name was Pauline Bosson (1894–1969). Although born in Chelsea, Massachusetts, she spent considerable time in Geneva, Switzerland, when she was a very young girl. She spoke French before she spoke English. When she finally went to elementary school in the Boston area, she was teased unmercifully for her poor English. For secondary education, Mother went to Miss May's School, where she met my future mother-in-law, Katharine Ayer Merrill.

Mother was a superbly complicated person. She was very strong and yet uncertain of herself. She worked hard for the Salem Hospital and for Grace Church in Salem. She was on all kinds of committees, taking on the tough chores that nobody else wanted. But ask her to chair a group, and she wouldn't want to do it. She didn't believe she could run a meeting, though those who knew her well would have said she could run a thousand.

She had lots of friends, people who adored her, but she could be very contrary. We learned as kids that if we had a choice of A or B and wanted B, we had only to say that we wanted A. She would argue with us for a while, whereupon we would say, "Okay, if you insist, we'll do B." In some ways, she was a pessimist. It was her view that summer was over on July 5. "It's downhill all the way from here on," she would say. Somehow that summed up her view of life.

My father, Robert Channing Seamans (1893–1968), was such a mild-mannered man that he seldom fought back when Mother became contentious. They had been married only a short while when Mother decided to splurge on their meager income and buy a roast of beef. She overcooked it, but when Father carved it, he characteristically said, "This is wonderful, Polly!" She said, "You know it's not!" Then she stuck a fork through it, took it out, and tossed it into the swill pail. Coming back into the dining room, she took her corset off and threw it at him, so annoyed was she by his forbearance.

My parents first met at a dance in Salem. As Grandfather Bosson was a director of the Naumkeag Mills, chaired by my great-uncle, Henry Benson, the Bensons invited Mother to visit them in Salem and to attend the dance. Mother said she didn't want to go, but Grandpa insisted: "Polly, it's important to me that you go." Later, Mother was invited back to the Bensons and went more willingly the second and subsequent times. Father knew another young woman named Ellie Rantoul, who had a sports roadster. According to Mother, Father used to tantalize her by driving by the Benson home with Ellie Rantoul when he knew Mother was visiting and might be looking out the window!

At the time of their meeting my father was a student at Harvard

College (class of 1916). He had been born in Marblehead, where my grandparents had the third house on the Neck—in the days when there was no paved causeway, and passage to the Neck depended on the tides. Like many families, they spent the summer in Marblehead and the winter in Salem. Grandma had wanted my father to be named Hugh Gerrish, an old family name. Aunt Rebe (pronounced "Reebee") Benson, my grandmother's sister and the wife of Henry Benson, often invited people to Sunday lunch at her home next door to my grandparents. One day shortly after my father was born, the Episcopal bishop was one of the invited guests. She called up my grandmother and said, "Carrie, the bishop's here. Why not have the baby baptized today?" Grandmother agreed and brought father over to Aunt Rebe's house. When the bishop asked the godmother, Aunt Rebe, for the child's name, she said Robert Channing, not Hugh Gerrish! Apparently Grandmother accepted the choice. I've always been very grateful to Aunt Rebe that my name is not Hugh Gerrish, though Channing was never a family name before that christening. As far as we know, it came straight out of the blue and into Aunt Rebe's head.

Grandmother Seamans, whose maiden name was Caroline Broadhead (1859-1949), was a real sport. On one occasion, when barnstorming pilots arrived at the Beverly airfield to take people for rides in their old open-air two-seaters, she climbed aboard in her long flowing dress and had a grand time. Her husband, my grandfather, Francis Augustus Seamans (1860–1931), ran Perrin Seamans and Company, a Boston hardware supply store.

Father went to Salem High School, Noble and Greenough, and Harvard College. He became interested in architecture as an undergraduate and wanted to continue his studies in this direction. Grandfather Seamans, who had never gone to college himself, figured a bachelor's degree was more than a man needed anyway, and he was certainly not about to send his son to graduate school! So Father became a certified public accountant.

Early on, he served as a bank examiner. About the time I became conscious of what he was doing, he was working in a brokerage house in Boston. After the stock market crash of 1929, however, he lost his job. The following year my father went back to work on a supposedly short-term assignment for the County Savings Bank, the bank founded by Grandfather Bosson. Grandpa had died, and Mother's older brother, my Uncle Campbell, had taken over the bank, as well as a lot of my grandfather's other affairs. My father offered assistance to people who could not meet their mortgage payments. It was a time when foreclosure didn't necessarily help a bank, because there was no market for repossessed houses. So beyond any humanitarian concerns, the bank had a real economic reason for helping people work out their mortgages. When economic conditions started to improve, home-made bottles of wine and all kinds of other tokens of appreciation began appearing in our house from people my father had helped.

At this time I was just starting to board at the Lenox School in western Massachusetts. Mother and Father came out to see me and said they weren't sure they could keep me in the school. Tuition, room, and board amounted to about $650, and the school was upping the amount to $750. While visiting me, Mother and Father went to see the headmaster to say they didn't see how they could possibly pay the additional $100. I found out afterwards that I had been allowed to stay on for a while at the $650 level.

My Brothers

I was born in Salem, Massachusetts, on October 30, 1918, the oldest of three boys. When I came home from boarding school for the first time at age thirteen, my brothers, Peter (born February 19, 1924) and Donald (born January 19, 1927), were waiting for me at the front door. I remember very distinctly being amazed at how young they were. Peter is five years younger than I, Donny eight years younger. We didn't have a great deal in common in those days, but we've since developed many common interests.

I got to know Peter better earlier because we were closer in age. The two of us weren't always very kind to our genial brother. If Donny wanted to join us, we would make up excuses to keep him out. For example, at Christmas time we would claim we were wrapping his presents. Donny went to Bowdoin College. Gene and I drove through Brunswick, Maine, where Bowdoin is located, not long after he had graduated. I knew Bowdoin had just changed presidents, so I went into a Western Union office and sent Donny a telegram signed by the new president. It stated that, in my new capacity, I wanted to review the status of all Bowdoin graduates to be sure that they came up to the standards of the institution. I requested that all graduates report to a local high school to take a series of tests and that Donny report to Marblehead High School for same. The woman at Western Union looked at my message and asked, "Are you really the new president?"

I answered, "Well, I'm not Abe Lincoln." The telegram was sent. Donny got terribly upset when he received it and started calling around. It took him a while to figure out that I had done the deed. He vowed he would never give me the satisfaction of knowing how upset he had been.

Peter and I ganged up on Donny until Donny got married, at which point his wife, Beverly, put an end to it (except for the final riposte from Brunswick, Maine!). About the second time we started teasing Donny in her presence, she lowered the boom. She's a very strong-minded person and a talented sculptor.

Boyhood

Like several of my boyhood friends, I was an electric-train buff. In the wintertime we would go up in different attics and lay down track and run locomotives all over the rafters.

In the summertime, starting when I was eight years old, I was sent off to Kingswood Camp in Bridgeton, Maine, where there were lots of sports. I enjoyed these, as I did the carpentry program, where we built model boats. Some of my friends back in Salem were very clever at making model biplanes, and I also joined in with them. We were all handy with tools. There was a weather vane over our house in Salem that I built when I was about nine years old. It didn't fall apart immediately.

As a boy, I was an avid reader of *Popular Mechanics,* and thus developed an early interest in machinery. With another friend of mine, I built crystal radio sets. By the time I was ten or eleven years old, I had figured out how to use one tube for amplification. With a headset I reached the point where I could actually distinguish stations. This was a great step forward. I couldn't always tell what the music was or what the speaker was saying, but at least I could tell if I was listening to music or to conversation! I used to hide under the covers of the bed at night and listen to radio programs, then pretend I was asleep if somebody came into the room.

I don't know where my interest in building things came from, though I do have at least one famous engineer in my lineage. Grandmother Seamans's maternal grandfather was Otis Tufts (1804–1869), who obtained one of the original patents for the elevator around the time of the Civil War. His model was steam driven. Right after the war, buildings in New York began to exceed five and six stories. His and similar inventions helped make that possible. The Prince of Wales is reported to have come across the ocean to ride in one of great-great-grandpa's elevators.

The aptitude in my immediate family that proved most useful to me in later life was mathematics. An accountant early in his career, my father loved numbers. As secretary of the race committee of the Eastern Yacht Club, he enjoyed calculating daily averages for all the boats. His only sibling, Uncle Dick Seamans, was one of the best accountants in the business, head of his own firm, Seamans, Stetson, and Tuttle, in Boston.

Tower School

From my earliest life in Salem, Massachusetts, I went to the Tower School, a private elementary school about five blocks from our Broad Street house. (Tower has since moved to Marblehead.) On my first day in kindergarten Ms. Runette and Ms Luscomb, who ran the school, took a look at me as I came in the door. "He's so tall," one of them said. "Start him in first grade." Consequently, I was always a little bit young for my class.

Tower provided a very good basis for going on to high school and college. It was imaginatively run, with much emphasis given to creativity over and above learning by rote. We made a lot of things with papier-mâché. We put on many plays, too, for which our poor mothers had to make all kinds of costumes and endure our acting.

The academic curriculum went in very strongly for Greek mythology. We studied Homer's *Odyssey* and other classics in great depth. We recited a lot of the material and acted out the stories. Classical Greece became almost contemporary in our minds. Tower also had quite a good French program, which would have served me better in later years if more had been spoken.

The public elementary schools in Salem did not have sports programs, but Tower did. We played football in the fall, hockey on a
freshwater pond in the winter, and baseball in the spring. We all kept
current on the great professional athletes of the time and tried to
pretend we were like them.

Mastoidectomy

Running counter to my love for sports was a history of sinus and res-
piratory problems. In the spring of 1926, while I was in the third
grade, I had measles. Just as I was recovering from measles, I got scarlet fever. A sign was tacked on the front of our house reading,
"Beware." We were quarantined! Father was allowed to go to work,
but Mother had to stay in the house with me.

After about a week I started having a problem with my left ear. A
specialist, Dr. Tolman, drained it. Then one afternoon Stumpy Reed,
handyman to Dr. Phippen, my regular doctor, began bringing a lot of
equipment into my bedroom. "What's going on?" I asked.

All he would say was: "I feel awful sorry for you, son."

When Stumpy left, I called Mother and asked what was up. Tears
came to her eyes. "I just wish it was me having surgery and not
you," she said.

I couldn't be taken to the hospital for surgery because I was still con-
tagious with scarlet fever. So Dr. Tolman, Dr. Phippen, and a couple of
their associates came and performed a mastoidectomy, the surgical
removal of infected bone behind my ear. A mastoidectomy seldom is
performed nowadays, because penicillin is so effective. But back then the
only thing a doctor could do was hammer away at the bone and remove
the infected area. In later years, my parents often recalled the sickening
sensation of sitting in the living room and hearing the sound of pound-
ing overhead. Finally they couldn't stand it any longer and climbed the
winding stairs to my room. From outside the door, they heard one of the
doctors say, "It's lucky we operated today. There wouldn't have been
any hope tomorrow." But I recovered without undue complications.

Marblehead

For the summer of 1932, just before I went off to boarding school, Father rented a house in Marblehead, where we went sailing and swimming. We had it for six weeks. The house did not have any hot water on the second floor. To take a bath, we heated water on the stove, carted it upstairs, and poured it in the tub.

The kitchen was an awful mess. I went over with my parents before we moved in and helped them paint it. The house grew with time and was eventually moved to a new site. My brother Donald lives in it today. It is almost unrecognizable from the days when my parents first rented it.

It was in Marblehead as a youth that I started sailing and playing a lot of tennis. Lots of Marblehead kids age ten and up sailed in the same kind of boat, a Brutal Beast. We would go charging around the harbor, racing and bumping into the big boats at anchor. We sailed and raced whenever we weren't playing tennis. David Ives and I eventually bought a slightly bigger T-boat. For the first month, we came in last in every race. We realized we must be improving the day we could still read the name on the stern of the next-to-last boat as it crossed the finish line. A great accomplishment! Before we were through, we won most of the cups that could be won. But that took several years' experience and a new set of sails.

Kent School

It was presumed that after the eighth grade at Tower, I would go to Kent, an Episcopal boarding school in Connecticut attended by my first cousin on my mother's side, Dave Bosson. He was quite a tiger, who did very well at Kent both academically and athletically. I also had another cousin, James Otis ("Jim") Seamans, who got straight As wherever he went. I was always being compared with the two of them and didn't measure up very well. This didn't bother me very much, but I could tell that it was of some concern to my parents.

Kent School had church services every day and was run by an Episcopal priest named Father Sill, whom I always picture in white robes slightly stained by tobacco. My father took me to see the school.

My uncle Campbell Bosson went along to introduce us to Father Sill. While visiting, I was put up in the infirmary along with some other applicants. We took exams in math, English, and Latin, then had physical examinations and interviews with Father Sill. I well remember climbing the stairs to his garret, arriving at the top, and seeing him at the other end of the room, smoking a pipe and looking very imposing in his white robes. He asked me a number of questions, including whether or not I had been confirmed in church. I said yes, having been confirmed just a few weeks before.

"What denomination were you confirmed in?," he asked.

As far as I was concerned, I went to the same Salem church as everybody else. I had never noticed what the denomination was. So I answered, "I'm not quite sure, but I think it begins with a P."

"Oh, you must be a Presbyterian."

I said, "No, I don't think that sounds right."

He hemmed and hawed a bit, then said, "You're not Episcopalian, are you?"

"That sounds right," I said. "I think that's what I am."

When I arrived home several days later, I found Mother in tears. She had just received a letter from Kent informing her that my application for admission had been rejected, although they had "enjoyed having Bobby there." I received a very low mark on the mathematics exam, but did better in English and Latin. The letter indicated that I seemed a bit immature, and I was encouraged to reapply the following year. When I told my parents about my encounter with Father Sill, they figured that was what had done it. This was no laughing matter to them. I'm sure they felt that the Seamans family had not measured up to the Bosson standard. What's more, my application to Kent had cost them $50, plus a donation for the new school chapel.

Lenox School

After hearing the bad news, my parents immediately whisked me off to the Lenox School in Lenox, Massachusetts, for an interview. Nobody ever asked me what school I wanted to attend. I had been sent to camp for four years, and nobody had asked for my views on that either. As a boy in that time and place, I didn't have much control over my destiny.

At Lenox, I met George Gardner Monks, the headmaster, and was given a couple of fairly perfunctory exams. In answer to one question about Greek mythology, I offered a long tale about Homer's *Odyssey* thanks to the Tower School curriculum. I was accepted on the spot! When I got back to Tower, I found that my friends had put up big signs saying, "Welcome, Bobby!" and "Congratulations on being admitted to Lenox School!"

Mr. Monks came from what was considered a top-drawer Boston family. He had gone to St. Mark's, one of the more prominent Boston-area boarding schools, and had become a minister. He was a great disciple of the Reverend William Greenough Thayer, headmaster at St. Mark's. Mr. Thayer felt that there was a need for a school that would admit boys who could not afford to go, for one reason or another, to St. Mark's, Groton, Middlesex, or the like, as well as boys who needed remedial training. The result, Lenox, was looked on as somewhat inferior.

It was a self-help school. The students did absolutely everything except cooking and laundry. We set the tables and washed the dishes. We swept the floors and mopped them. We had "work holidays" on some Saturdays, when we boys might have wanted to be out doing other things. All day we raked and generally cleaned up the grounds. As a result of my Lenox education, I can wash a dish better than anybody I know. After my wedding, I made the mistake of demonstrating my wife's relative inefficiency in the kitchen by showing her how to wash dishes. Served me right—I ended up washing them for a good stretch!

School began in mid-September, and we all stayed on campus until December 20, except when parents came to take us out to a meal. (In those days there wasn't a Massachusetts Turnpike, and it was a bit of a struggle for my parents, with two younger sons in tow, to get out there.) For more than three months it was no weekends, no Thanksgiving, no going home. I remember thinking in September that I was going to spend the next fourteen weeks—roughly 1/300th of my whole life span—incarcerated! Still, Lenox was a remarkable school, and it gave me a wonderful background. I had some very good masters during my five years there.

A student tends to get a reputation in a given area, which becomes very hard to change. I got a reputation for not being too facile in English. Mr. ("Snark") Clark, the head of the English department, was always calling me in and forcing me to rewrite papers. This was a big drag, but very good for me. Nor was I the most thoughtful student in Sacred Studies. My parents were called in once during my sophomore year because I had been caught playing cards during a Sacred Studies exam. I had done no work in the course, and I could see no conceivable purpose for taking the exam. Why pretend that I had done work I hadn't done? My parents were extremely upset with me and made this quite clear by making the long drive, four hours each way, to confront me. I reformed to some extent. Before the year was over, I got Sacred Studies up to a passing grade.

I really enjoyed math and sciences, which came easily. I did well enough in mathematics my sophomore year that Mr. Monks, who taught the course, thought it would be a good thing, as an experiment, for me to take the college boards in math two years early. So that I could do so, I stayed in school one extra week and went over the junior and senior math curricula. As soon as I left the exam room, I knew that I had made one stupid mistake in proving a geometry theorem. I got an 85; if not for that mistake, I would have had 100. After that, the head of the math department gave me and one other boy special problems because we had already fulfilled the school's math requirement.

The Lenox subject that I found most interesting was senior physics, also taught by Mr. Monks, who was a great academic inspiration, especially in my area of greatest interest. As headmaster and minister, he also had an influence on my sense of values. One time, I came in from football and was the last person to leave the locker room, which was in the basement of the main building. I noticed a big electric switch, and just for the fun of it I pulled it. It turned off all the lights in the school. Subsequently, there was an inquiry to determine who had perpetrated the deed. We students were all assembled in the dining hall and were told we were going to stay there until the culprit owned up. I confessed and was given my punishment. The earthworks dam that contained the water for our natural ice-hockey rinks tended to leak. I was tasked with digging a trench all the way around the dam and filling it with clay in order to stop the leaking. I spent several weekends at the dam accomplishing this. What did Mr. Monks have to do with all this? Remarkably, he came down to the dam and worked alongside me.

Harvard College

For a long time I maintained that I wanted to go to the University of Wisconsin after graduating from Lenox. I guess I wanted to go somewhere other than Harvard College, where my two uncles (Richard Seamans and Campbell Bosson) and my father had studied. Why Wisconsin? It was reasonably far from home. In the end, I applied to only one college, Harvard, and began studying there in September 1936.

I had received a high enough grade in English on the college boards that I was allowed to skip the standard freshman English program at Harvard, a noncredit writing course. I was delighted my cousin Jim Seamans, who had been an excellent student at Exeter and who had entered Harvard the same year I did, hadn't done as well on the exam and was required to take the course. Everybody in the Seamans family was convinced that somehow the two examinations had been swapped! What really happened was this. The college boards in those days weren't the multiple-choice, true-or-false affairs of today. The questions were substantive and required essay-type answers. On the English exam, we were asked to discuss books of four different kinds. With my background in Sacred Studies, I selected four books of the Bible. This staggered the examiners, or so I was told afterwards by "Snark" Clark, who knew the teacher who had read my paper.

The training I had received at boarding school was so much better than that given in most high schools that my freshman year at Harvard was straightforward. My roommate, Stratton Christensen, and I worked hard and got good grades. I was reasonably straitlaced, studying at night and going to classes in the day, while trying to compete at football, hockey, and baseball. I didn't do very well in college athletics, but I did get out there and play the game. The summer after freshman year I didn't have anything to do, so I took a full-credit course in surveying given at Squam Lake in New Hampshire. I found it fascinating. I got to know Professor Albert Haertline quite well, the camp director and a Harvard classmate of my father. He was an old-time civil engineering type, who wore knickers and big boots and liked to hike up and down through the woods, surveying. I also got to know a number of students who were already concentrating in engineering. I pursued engineering studies from there on.

The engineering faculty at Harvard was fairly small, but it had some remarkably good people. Professor Haertline became my advisor and was most helpful in his discussions with me. Den Hartog, who taught mechanical engineering, became a captain in the Navy during World War II, then came back and headed up the department of mechanical engineering at MIT. The extent to which a Harvard engineering student could concentrate in a particular area was limited. For me, there was beauty in that. I didn't have a very clear picture of what I was going to do with my life. It had been expected of me that I would go to college and get a degree, and I was doing so. But what I was going to do with the degree I didn't know. So Harvard's generalist approach to engineering allowed me to sample different fields—electrical, mechanical, civil, and aeronautical engineering. I discovered that I was especially interested in aeronautics, as taught by Bill Bollay. He wasn't a charismatic figure, but he had considerable technical ability. After serving as a naval captain in World War II, he went out west and became one of the key people in the Autonetics Division at North American Aviation.

With extra credit for the summer surveying program and for the advanced English course I had taken in place of the freshman noncredit writing course, I entered my sophomore year with six of the sixteen credits needed to graduate. So I thought, why hang around? Why not polish it off in three years? That meant taking five full courses each of my last two years. Against everyone's advice, I took six courses during the second half of sophomore year. One of the six was Astronomy 2a, which included a section on celestial navigation. In one of the key assignments, we were put in a hypothetical boat off Bermuda, given a lot of star sights, and asked to determine where we would make landfall. The night before the exam I still had not done the assignment. Nor had I studied for the exam. I had learned that the best way for me to study under that kind of last-minute pressure was to work for two or three hours, catch a couple of hours' sleep, then get up and work again—alternating work and sleep throughout the night. This time, however, I was in such deep trouble that I got hardly any sleep at all. The following morning I was so afraid that I would fall asleep during the exam that I filled an ink bottle with whiskey and drank some periodically during the exam to stay awake. Not very smart.

It was during this already overworked sophomore year that I realized there was more to life than books. There was wine. There were women. There was song. As undergraduates we were invited to too many dances, and I decided I was going to go to every one I was invited to and have a good time at each. During the winter months I also went north to ski almost every weekend. In general, my life became a pretty big mess. I got very little sleep. More than once, I went into class wearing my tuxedo from the night before. Then in February I got a terrible cold. When I finally got around to seeing the doctor at the college infirmary, he sent me to bed for ten days. The day I got out of bed, there was a duplicate bridge tournament I wanted very much to play in. I played until three in the morning and never felt more tired in my life. I did pass all six courses that spring semester, but I didn't do particularly well in any of them.

Rheumatic Fever

During that summer of 1938, my pace never let up. Right after exams, I went down to New London for the crew races, then raced back to Boston in a friend's sailboat. I arrived back in Salem just in time to be an usher at my cousin's wedding. Two days later, I was aboard the good ship *Columbus* of the German-American Line, sailing for Plymouth, England, with my friends David Ives and Johnny Brooks. The itinerary called for me to meet my parents and brothers on the Continent after a bicycle tour of England with Dave and Johnny. The three of us didn't live a very healthy life on shipboard. We had almost the cheapest room on the boat, sleeping in a tripledecker below the waterline. It was so horrible in third class that we found a trick way into first class and from there into tourist class. The stewards in first class were always rather suspicious of us when they came to serve tea at five o'clock, but we stuck close to some girls who were traveling in first class with their parents, so the stewards couldn't kick us out. We met a second set of girls, who also provided cover for us, in tourist class. That was a lot of fun.

As soon as we got to Plymouth, we hopped on rented bicycles and went tearing off over hilly country, ending up in a little town called Launcestown. That night I was as dizzy as I had ever been and felt terrible. The following morning I went to see a doctor who said, "Obviously you've had a strenuous crossing. Get a couple days' rest before you continue your travels." My condition persisted. After completing my itinerary in Great Britain earlier than planned, I parted company with my friends and headed for Berlin where I joined my Grandmother Bosson,[1] Mother, my brother Peter, my aunt and uncle, and some Bosson cousins. To get there, I flew across the English Channel, my first flight in a plane. Then I had a couple of days in Paris on my own, before getting on the Berlin train. I got in a taxi at the Berlin station and arrived at the Hotel Bristol without enough money to pay my taxi fare. Fortunately, someone in the family was there to bail me out.

I told my mother about my poor physical condition, and she insisted that I go to a doctor in Berlin. He told my mother, "Oh, he's fine. All this boy needs is some mountain climbing." Later that summer in Switzerland, where Father and Donny joined us, we did do some climbing. On one expedition, while roped together on a snowfield on our way up the Breithorn, I said that there was something very wrong. A guide took me back to the hotel, where my condition did not improve.

We went on to Italy where we spent two weeks on the Mediterranean. Then we sailed home aboard the Italian liner *Saturnia*. By the time we got back to Marblehead, the exhaustion I felt was indescribable. I slept all night and woke up feeling worse than when I went to bed. I went to see Dr. Cadis Phipps. He said that I had endocarditis, whatever that was, and that I should take it easy. Under his care, I found my third year at college very different from the other two. I could not exercise. If I went out and rowed a short distance, I felt weird. I continued to go north on weekends with my friends, but I couldn't ski. When I tried to, I got a very unpleasant disembodied sensation. So I took pictures instead. On the positive side, I spent more time studying than I had during my sophomore year. As a result I finished up my final year at Harvard in 1939 getting good grades. I missed graduating with honors, but I did manage to complete my studies within the three-year period I had set myself.

---

[1] Once, when sightseeing with us, Grandmother made a memorable pronouncement. While viewing a monument in what is now Innsbruck, Austria, one of us asked her what it was. She answered, "I don't know what it is, children, but it's very important!"

In the summer of 1939, following my third and final year at Harvard, I took a bus to a ranch in Montana. Once there, I knew I was in trouble and called Dr. Phipps. "You had better come home," he said. When I went to see him in Boston, he talked of endocarditis again and used other terms that didn't have any meaning for me or my parents. Then he said, "I think you had better have some real bed rest." I was put in bed for an indeterminate period of time.

This had quite an impact on how I viewed things. As far as I was concerned, I had had it. My life was going to be severely limited from then on. I would not be able to go mountain climbing or ski or play tennis ever again. People were very generous of their time, coming to see me and doing many thoughtful things for me, so that I got to know a lot of people better than before. I also did a lot of reading that I otherwise wouldn't have done. I had several friends in the medical profession who brought me material to read. I started thinking of shifting from engineering into medicine, assuming I would eventually be up and around again.

My parents refused to be disheartened. They set about finding the best cardiologist available. Dr. Phipps was a good doctor, but by this point we knew he wasn't our doctor, because he couldn't explain to us in a helpful way what was going on. After a lot of investigation my parents found Dr. Paul Dudley White, one of the great cardiologists. When the King of Sweden had a heart problem, he called on Paul White. But how to get him to come from Boston to Salem to see me? By sheer coincidence our next-door neighbor, Dick Wiswall, had been Dr. White's roommate at Middlesex School. Through him we got a house call.

I asked my parents not to tell me what day Dr. White was coming, because I was sure that it would worsen my condition by making me apprehensive. I wanted him to see me under the best possible conditions.

On the walls of my room, I had pictures I had taken of friends skiing and sailing. When Dr. White came in at long last, he began by commenting on them. I responded by saying, "I guess I'll never be able to do that sort of thing again."

"Well, how do we know?" he said. "Let's take a look." He took out his stethoscope and listened. "That's a really peppy-sounding heart," he said. He had some tests run. Soon he had diagnosed my condition as rheumatic fever. He did not prescribe any pharmaceutical or surgical remedy but rather encouraged me to slowly get myself up and about.[2] Dr. White's optimism, combined with my confidence in him, changed my life. Soon after that visit, I was able to get up ten minutes a day, then fifteen minutes, then twenty. After a month I was able to go up and down stairs once in a day. It was not long before I could walk out the front door, then walk to the Wiswalls and back. I finally reached the amazing day when I walked all the way around the block by myself. What a wonderful feeling: I hadn't had so much freedom, exercise, and fresh air in months!

After a half-year of recuperation I was mobile again, and I went to the graduation exercises of my Harvard class in June 1940. By early in the summer, I seemed pretty well cured, and I found I could be quite active again, with Dr. White's blessing. I played a little bit of tennis. I was able to sail with no ill effects. Every time I tried something new and got away with it, I thought, "This is terrific. I never really thought I was going to be able to do this again." I went back to Harvard that summer to take courses in biology and botany, prerequisites if I was going to pursue my new interest in medicine. It was a pleasant experience. I particularly enjoyed the course in biology.

Gene

I had other pleasant experiences that summer, all stemming from an encounter the previous fall, while I was recuperating in bed. I had received an invitation to the coming-out party of Rosemary Merrill, who summered at her parents' Prides Crossing home known as Avalon. Mother, who had known Mrs. Merrill since their days together at Miss May's School, called for me to say that I would be unable to attend. Not long after that, I got word that Mrs. Merrill was coming over to see me. Little did I know that a very beautiful nineteen-year-old by the name of Rosemary was coming along. Romey, as she was called, was very striking—with black hair, a lovely face, and a lot of bounce. Within the next few months, she visited me two or three

---

[2] Years later, I ran into White's assistant, Dr. Conger Williams, who showed me the file cards White had kept on my case. "We weren't really sure what to do with you," he confessed. Evidently Dr. White almost decided to put me in a hospital and run catheters into my heart to determine what was going on. In those days that would have been a pretty extreme measure, although it's common practice today.

more times. Her family returned to their winter home in Washington, D.C., and we wrote each other a number of times.

The following summer I thought I would go down and see where the Merrills lived. Anyone else would have called ahead, but I didn't. I was driving around in the area in my old Dodge and headed unannounced up the Avalon driveway. It was considerably longer than I had expected, and the house was bigger. Adolfo, the butler, came to the door and asked what I was doing. I said I was there to see Miss Rosemary Merrill. He let me in. Romey was notified but didn't recognize me when she came down the big circular staircase. She had never seen me standing up! Mrs. Merrill was alerted and very cordially invited me to sit down in the extensive living room. Eugenia, Romey's younger sister, who was known as Gene, was called in. She was dressed in a paint-spattered smock. She had her own studio in the house, devoted exclusively to her painting. As we sat there chatting, I didn't have the finesse to excuse myself after a moderate time. So what else could Mrs. Merrill do? She asked if I would like to stay for lunch.

"That would be wonderful!," I said.

Unfortunately, Romey had to leave to go racing in Marblehead; so I had lunch with Gene and Mr. and Mrs. Merrill. Afterwards, I was asked to join them on the porch.

"Sure!" I said. They asked if I had ever played Chinese checkers. I had; so we had a game and a cup of coffee. Finally I did leave, never realizing how displeased Gene had been to have left her paints in deference to an uninvited stranger.

Mother was very curious about such social encounters, and when I got home she asked me about my visit. Didn't Romey have a sister? Had I met her? What was she like? I said, "Well, she's a little bit plump and quite young." I really didn't have a very clear impression of Gene after that first visit, but when I was invited to Avalon on a couple of occasions later that summer, I came to realize that my first observations of her were inaccurate, to say the least.

Postgraduate Studies

When September 1940 rolled around, I had to make a decision on my future. I talked to my old undergraduate advisor, Professor Haertline, a wonderful sounding board. Though he didn't tell me what to do, he pointed out that my undergraduate engineering studies were a very good background for medicine, if I wanted to follow that route. Then he asked me, "What kind of a doctor do you think you would be?"

"Well, to be honest with you," I answered, "I think I could be a better engineer than doctor." I realized I had taken naturally to engineering sciences during my undergraduate career and had had no difficulty getting top grades in the field. That seemed to be an indication of how I might make out professionally. Biology had not come as easily for me. The subject matter didn't fit as neatly in my mind. For me there was an intellectual difference between medicine and engineering. As a doctor, I'm sure I would have liked working with people and trying to figure out what their problems were. But it appeared to me then (and still appears to me now) that there's a lot more mathematics in engineering, that medical practice is much more a question of remembering things—anatomy, symptoms, prescriptions, and so on. I'm sure a practitioner has to be fairly nimble mentally. But doctors are often writing a prescription they've memorized for a set of symptoms they've also memorized.

The upshot of my meeting with Haertline was that I decided to go back to Harvard to do graduate work in aeronautical engineering under Bill Bollay. I was not quite sure that I was physically ready for a full load, so I signed up for just two graduate courses that fall. A good friend and classmate of mine, Richard E. ("Dick") Lewis, was entering the same program. The day after the semester began, he said, "I wonder if we're doing the right thing. If we're really interested in this field, shouldn't we be at MIT?"

By this time I had realized that Professor Bollay's course in aeronautical engineering was not very different from what we had taken as undergraduates. The other course I had signed up for was being given Fridays and Saturdays by an adjunct professor coming up from the Chance-Vought Corporation in Stamford, Connecticut. Dick quite rightly said, "Look, if you go over to MIT, you're going to have full-time faculty members, not adjunct professors." Then he said, "I'm going over to MIT this afternoon to see the dean of admissions. Do you want to come along?"

I knew very little about the institute, but I said, "Sure, why not?" So we went over to Dean B. Alden Thresher's office.

"Okay, boys," he said, "tell me a little bit about yourselves." When we were finished with our brief oral résumés, he said, "You two did quite well at Harvard. I think we can get you started here as sophomores, although you may have a few deficiencies."

We both stood up and said we weren't interested in starting all over again towards a bachelor's degree. "Well," he said, "maybe I'm underestimating the extent of your engineering background. Here are some forms you can fill out. Take them around to the various departments and see how much credit they'll give you."

When we had left the dean's office, Dick said, "I'm going to stay at Harvard."[3] I took my form home, read it, and filled it out. Then I read the MIT catalog and went around to the departments, as Dean Thresher had recommended. I did some bargaining over equivalencies and quickly found that I was well beyond the sophomore level. My last stop was at the office of Professor R. H. Smith, acting head of what was then called aeronautical engineering (now aeronautics and astronautics). By this point, I was beginning to visualize myself taking one undergraduate year at MIT. I liked what I saw, walking up and down the corridors: all those labs and machine shops! Compared to Pierce Hall, the single building devoted to engineering at Harvard, MIT had so many interesting alternatives and possibilities.

Smith said, "I just can't understand why you want to be an undergraduate. Why don't you come here as a graduate student?" That sounded fine to me. "Let me call Thresher," he said, picking up the phone.

He and the dean had quite an argument on the phone. At one point Smith turned to me and asked, "Have you had descriptive geometry?"

"Sure, I've had it," I answered.

When he finally hung up, Smith said, "We can admit you to graduate studies in aeronautical engineering." I later learned that it is the prerogative of each department to admit whomever they wish to their graduate school.

"Well, all right," I said, "but now, how long do you think it will take me to get a master's degree? You've seen my record."

---

[3] Dick Lewis received his master's degree from Harvard in 1941, designed aircraft for Chance Vought during World War II, and had a successful career in the aircraft industry.

Smith said, "You haven't had very much aeronautical engineering, so you're going to have to do a lot of work...unless, that is, you decide to specialize in instrumentation."

"What's that?"

"I'm not sure I can explain it well. You ought to go and see Dr. Draper. That's his province."

I was pretty excited about what I had seen. I went home and told my parents I was going to MIT. They were quite upset at my vacillation. First I had talked about going to medical school; then I was going to graduate school in engineering at Harvard; now suddenly I'm going to MIT! Classes started the following week, and all of a sudden I was taking a full set of courses. I did well. I had a cumulative average of 4.8 (on a scale of 0 to 5). I not only enjoyed the courses, I worked hard at them and didn't fool around with a lot of other things.

Doc Draper

Dr. Charles Stark ("Doc") Draper, the instrumentation professor, was a Stanford University graduate with a bachelor's degree in psychology. After Stanford, he came through Cambridge on a lark, stopped at MIT, and decided he wanted to study there. Before he was through, he had taken more MIT courses for credit than anybody has ever taken before or since, earning a B.S. in electrochemical engineering in 1926, an M.S. without specification of department in 1928, and an Sc.D. in physics in 1938. He became especially interested in propulsion and engines. He found that there were no reliable instruments to tell a research engineer, much less a pilot, how an engine was performing. A pilot himself, Doc became increasingly interested in gyroscopes, horizon indicators, and the like. Then he moved into the field of fire control, the science of aiming guns at moving targets.

I didn't get a chance to see Doc Draper before my first day of classes at MIT, but I did attend the first lecture in his introductory course in instrumentation, number 16.41. (Everything at the institute is by the numbers!) It was standing room only in the largest classroom in the building. Doc Draper came bouncing in wearing a green eyeshade. He was quite short of stature and heavyset, but I could see right away that he was one of the most dynamic people I had ever met. He had a broken nose because, I later learned, he had once been a prizefighter.

"I'm delighted you're all here," he said, "but let's not kid ourselves. If you're going into this profession, you're never going to make a lot of money. Face it right now. You'll be lucky to earn $10,000 a year. You're not going to have money for race horses or for buying a mink coat for your wife. But," he added, "you'll have a hell of a lot of fun."

At that point $10,000 seemed like a lot of money to me. But that was secondary. I was intrigued immediately by the material. The discussions were lively, with a lot of give-and-take. Doc Draper was running MIT's Instrumentation Laboratory, and he brought material from his lab work into the classroom. I quickly realized that I was faced not with somebody who had never practiced but with a person who was leading the pack and telling you what he was doing. I reacted differently to him than I had to other, more theoretically oriented professors. Doc Draper was making a living doing something important! All I wanted was to know more about it myself.

The Instrumentation Lab was working with the Sperry Gyroscope Company on vibration-measuring equipment, so we students were soon experimenting with accelerometers and vibration pickups and all the electronics connected with them. The lab was also doing classified work (which we students had nothing to do with) on anti-aircraft fire control for the British navy. Why the British navy? Because the Sperry Corporation, which manufactured the U.S. Navy gunsights then in use, had convinced our military that their system—much more complicated (and therefore more expensive!) than Doc's system—was the only option. After Pearl Harbor (December 7, 1941), the government commandeered all foreign research, and Doc Draper's British navy work was turned over to our military by fiat.

That first year of graduate studies at MIT was far and away the most exciting academic year I had ever had. For example, I took a course in Laplace transforms and spent a whole weekend solving problems using the theory of complex variables. Suddenly I understood this type of mathematics, and the power of it was quite a revelation to me. To think of the problems it could solve! But nothing could match the excitement of instrumentation. The real fun of it was to do the studies, then actually build the hardware based on the theory. And sometimes it worked! Doc Draper would describe the fun of making something work as "defeating Mother Nature." Design something, build it, and test it successfully—and you've made something useful for a small part of our society.

It was the first time in my life that I could begin to visualize what it would be like to go into a field as a professional rather than just to study it in books. I was becoming very intrigued with airplanes—what permitted them to fly efficiently, what powered them, and what controlled them. Before the year was over, Doc Draper had taken us through all the major instruments that are required to pilot, to navigate, and to test a plane, and I could clearly visualize the possibility of working either for an aircraft or instrumentation company or for the National Advisory Committee for Aeronautics (NACA). The whole business was fascinating.

In Love

I didn't miss watching any varsity games while I was at Harvard. During the fall of 1940, while I was beginning my work at MIT, I invited three young women to go to three different games with me. While at a party at Avalon, I invited Gene Merrill. She didn't answer yea or nay right away, but as I was leaving through the foyer, I felt a hand on my shoulder. It was Gene, saying she would like to go to the game. I had a wonderful time with her that Saturday afternoon. We drove to Cambridge in my old beaten-up Dodge coupe and parked near Leverett House. After the game, we came out to the car and found a window smashed. Several things had been stolen, but fortunately Gene's fur coat was not one of them. It was a cold night, and on the drive back to Avalon she had to sit a little bit closer to me than she otherwise might have.

Before Christmas 1940, I was invited to Gene's coming-out party in Washington. By then I had developed a strong interest in her, and I was disappointed by my three days there. I found her a bit distant. I came up with the idea of taking a trip to Annapolis in order to have an opportunity to be alone with Gene. My cousin Jim Seamans lived there, and I thought that the hour's drive back and forth would give us a chance to talk. When I broached the subject, Gene said, "I've talked to Mother, and she said take the station wagon"—meaning I was to go alone! By the time Appel, the Merrills' chauffeur, drove me to the station a couple of days later, I was not sure I would be seeing Gene again.

When I got back to MIT, I was immersed immediately in preparing for exams, so I was able to take my mind off Gene. I knew Tina Vaughan, one of Gene's great friends, and to my surprise, I found that Tina was quite encouraging. At least, she said all was not lost with Gene. I was pleased. Then I got a couple of unsigned valentines. One had been mailed in Boston. When I was with Tina not long after that, I was delighted to learn that Gene had been in Boston at the time of the postmark.

I had done quite a bit of sailing with Gene's and Romey's older brother, Keith, in summers prior to 1941, but that year I found myself being invited to race with him more than ever. The Merrills were early risers, while we Seamanses were just the opposite. I got phone calls at quarter to seven, when everybody in our house was sound asleep. It was Keith asking if I could go racing with him. When we met, lo and behold, the crew would be Keith, Gene, and myself. I invited her to a few dances, and we got to know one another quite well.

Once we went with Keith and another young woman to the Ritz in Boston to hear Ruby Newman and his orchestra. The music stopped at midnight. When Gene and I got in my car to drive home, I asked her if she would be at all interested in going over to MIT to see the laboratory where I was working. I simply wanted to stretch out what was for me a most enjoyable evening. By the time we got to Avalon, it was about 2:30 a.m. I began walking Gene to the front door, when we both saw the silhouette of Mrs. Merrill in her bathrobe, standing in the doorway. Gene surreptitiously put her hand behind her back and gestured to me to back off and get going. A couple of days later, when I next saw Gene, she told me that her father had grilled her on the layout and contents of my MIT lab. Apparently Gene satisfied him by talking about the stroboscope and other equipment. Another time, when we were out bicycling in Topsfield a bit later in the day than Mr. Merrill thought appropriate, she was interrogated again.

I realized by late in the summer that things were developing with which I had never been involved before. I had never known a young woman so well, nor had I ever had such a good time with one. I do not know to what extent Gene or I was taking the lead. I guess it was mutual.

Master's Thesis

One day that summer I dropped by MIT to take a look at what I might do for a thesis in the fall. I went into Doc Draper's office and asked him if he had any suggestions. He talked about the program he was working on for Sperry Gyroscope Company, developing vibration-measuring equipment. He said that if I would care to work on it, he would take me on as a research assistant; I could satisfy my thesis requirement; and he would pay me, to boot. I was going to get paid to study! It sounded like a great idea.

"When do I start?," I asked.

"It would be great if you could start right away." I was assigned to Professor Homer Oldfield, whom everyone called Barney, after a famous racetrack driver of an earlier era. I spent my days in the basement of the building testing the performance of vibration-measuring equipment. Just outside my door was the place where all the trash in the building was collected. This work environment wasn't very attractive or aromatic, but it was satisfying getting some good, hard data and checking it against theory. The people in the lab, together with the faculty, were a great group with which to work. Early on, Doc Draper sent me over to three of his friends in electrical engineering: Edgerton, Greer, and Germashausen. They turned out to be the founders of EG&G, a very successful corporation. The scope of what was going on at MIT was at least an order of magnitude greater than what I had experienced in engineering at Harvard.

Soon, to my amazement, Barney Oldfield wasn't around very much. He seemed to be in and out, mostly out, leaving me alone with the equipment in the basement. He probably knew he was about to be called up into the Army Signal Corps and was straightening out his affairs. About six weeks after my arrival, Barney was in the Army. Doc Draper came to me and asked if I would be interested in teaching in Barney's place. I said I would. As fall rolled around, I found myself an instructor at MIT! That first semester, I helped teach 16.41, the basic instrumentation course. I began by grading papers and setting up the lab experiments to be carried out by the students. By the following spring I was in front of a class, doing some of the teaching myself.

My master's thesis, which I completed the following year, had the unlikely title of "Design and Test of a Vibration Pickup with Improved Performance by Hydrodynamic Effects." That may not mean an awful lot to the layman, but to me it meant a summer job in Pasadena immediately after Gene and I were married.

Engagement to Gene

My professional start-up with Doc Draper overlapped my start-up with Gene. In November 1941, just as I was getting established as an instructor at MIT, Gene came up from Washington to go to the Harvard-Yale game with me. She stayed with my family at a house my parents had rented from friends in Marblehead. We had a lot of things planned for the weekend, but at one point early on, we drove down to Revere Beach, stopped in the big parking lot there, and looked out at the ocean. I led up to the big question gently. "I want to ask you something," I said, "and I would like you to think about it before answering." I was afraid I was going to get the wrong answer, namely no, and I thought that I would like at least to enjoy the weekend.

Gene later told me that she had had a very hard time not saying yes right off the bat. Without giving me an answer, she got back on the train for Washington with her mother and with Romey, who had come north to visit the Merrills' Prides Crossing neighbor Caleb Loring, Jr. On the way to Washington, Romey told Mrs. Merrill that Caleb had proposed to her. Gene couldn't resist telling her mother that I had proposed on the same day. A weekend or two later Gene came back for another visit and accepted my proposal. (Caleb got the same answer from Romey.) Gene and I went out for a bike ride that weekend. When we got back, we heard on the radio that the Japanese had attacked Pearl Harbor.

Caleb and I were invited to the Hillsboro Club in Pompano Beach, Florida, to spend Christmas with the Merrills. I couldn't imagine not being home for Christmas; I had always had Christmas dinner with my great aunt Rebe Benson and about fifty or sixty other family members. So I told Gene I would leave Boston on Christmas afternoon and be at Hillsboro over New Year's. During the previous month I had been very busy and had not written many letters. I arrived in Florida, not knowing that Gene had told her family about our intentions. On the second or third day Gene said, "Don't you think it's time you had a chat with Mother and Father?"

"What's the big rush?," I asked. Finally one evening I met with Mr. and Mrs. Merrill. Mr. Merrill started in by asking, "How do you think you could possibly provide for my daughter?"

I thought I was pretty well-off. After all, I had a job that was paying me $1,200 a year. I said, "I believe we can work it out."

"Maybe you can, but I just want you to know what it's like to have a daughter who is very fond of somebody, a daughter who looks through the incoming mail every day and who finds nothing there. Do you realize the way you've treated her?" He went on in this vein at considerable length. It was not a very happy encounter.

I started to become afraid of what I might say. I finally said, "I don't think this meeting is going anywhere," and got up and walked out the door. Gene could tell something was wrong the moment I rejoined her. We walked about a hundred yards down the beach, and I became violently sick to my stomach.

Wonderfully, Mrs. Merrill stepped in as intermediary and arranged another meeting of the four of us together. We had a lovely lunch the following day, and I was elated by the outcome.

During my visit, Gene and I used her painting as a pretext for getting away alone together. After I had driven Gene on three or four consecutive days of artistic expeditions, her mother very sweetly said, "Gene, I think it's about time you showed us what you've been painting." There was nothing to show, but Gene did a very quick and skillful job on a fresh canvas and managed to pass it off as her work for the week. I still treasure that picture.

That spring my parents had a party for Gene, so that all of my friends in Marblehead and Salem could meet her. Mr. and Mrs. Merrill came north with Gene and stayed with us for a few days in Marblehead. Father and Mr. Merrill got along very well. At one point Mr. Merrill said, "Bob, your son probably figures that I've been pretty rough on him, but I think that young men these days have it much too easy. A fellow has to realize that when he takes on responsibility for someone for the rest of his life, it isn't something to be done casually." I found out later that Mrs. Merrill's father, Frederick Ayer, had given Mr. Merrill somewhat the same treatment. Mr. Merrill had been forced to wait a couple of years before marrying Gene's mother.

We formally announced our engagement on March 22. I drove

down to Washington for the occasion with my parents. I kept telling Mother how wonderful the trip would be: March in Washington! Mother, who was somewhat psychic and often pessimistic, kept saying, "I don't think this weather looks too great."

"Look," I said, "the flowers are coming out."

Mother wasn't convinced. We got to Washington on Friday. Mother woke up at three o'clock Saturday morning and said to my father, "I think it's snowing out."

"Come on, Polly," he said, "go back to sleep."

By the end of the blizzard, Washington had eighteen inches of snow. For as long as she lived, Mother never let me forget. As a result of the storm, a lot of people called to say they couldn't make it to the party, and we spent the whole morning shoveling snow from the Merrills' driveway. It was a wonderful party anyway.

World War II

Gene and I came together during a highly volatile period for our country. Immediately after Pearl Harbor, people were signing up for military service in droves. This marked a huge turnaround in public sentiment within a year.

Many of us in the Harvard class of 1940 had taken Government A with Professor Holcomb during our freshman year (1936–1937). I remember Holcomb saying, "Look to your left, boys; look to your right. One of the three of you won't be here when you graduate." He was right. We graduated two-thirds of those who started. He also said, "Get out those brass buttons and shine them up. You're all going to be in the military before you know it." This was met with almost universal disbelief. It didn't bother any of us as much as the thought that we might not graduate.

Even after Hitler began storming across Europe, the feeling on campus was: "We're not going to get involved in this mess." At our graduation ceremonies in June 1940, a graduate from the class of 1915 talked about World War I—how his generation had done its share and how it now was up to our generation to do its share. We shouted back, "You're not going to see us going over there and getting killed!"

During the summer of 1940, before graduate school, I asked Dr.

White, my cardiologist, whether he thought I could go into the Navy. "Well," he said, "it wouldn't hurt to try." So I went around to see if I could enter the Navy's V-7 program for college graduates. I didn't pass the physical. Curiously enough, my rejection had nothing to do with my recent rheumatic fever. I got tripped up on the color-blind test. I couldn't believe it. "I'm not color-blind," I said.

The doctor said, "According to these charts you are."

I got my parents to arrange an appointment with an eye doctor, Dr. Dunphy. I told him my sad story, and he tested me. "You are color-blind," he said. So I did not have the chance to enlist.

Later on, everybody who could walk a straight line was drafted, but by then I was an instructor and involved with classified work for the armed services. MIT rank-ordered all employees of draft age, according to who was needed most to help with the war effort. The Selective Service used this to determine who would be more valuable out of uniform. MIT had me in the top echelon of those who were of draft age. As a result, I was never drafted. Later in the war, when I got to know an Air Force colonel while working on a gunsight for fighter aircraft, I explored with him the possibility of getting into the Air Force. "No," he said, "you're doing more good staying at MIT."

It was not an easy time. My brother Peter and my brother-in-law Caleb Loring were in the Navy. My cousin Jim Seamans transferred from Harvard to Annapolis, where he graduated in January 1942. Two weeks after graduation he was aboard the destroyer Truxton when it rammed into rocks off Newfoundland. He was one of two officers who survived. Covered with diesel fuel, he swam for shore in a blinding blizzard and high seas. As he got near land, he heard somebody say, "I'll be right there, Bud," and then he blacked out. He woke up in somebody's farmhouse with frostbitten feet and hands.

The impact of the war was everywhere, and we were inundated with thousands of such stories. In March 1942, after my engagement to Gene had been formally announced in Washington, my college roommate Strat Christensen rode back north with me on the train as far as Philadelphia. I went on to Boston. By the time of our June wedding, Strat's ship had been sunk by a German submarine, leaving only two survivors. I would have preferred to serve in the military; however, I had the satisfaction of performing several jobs for the armed services—

developing equipment and installing it on an aircraft carrier when under way, and flying with the Air Force to help test fire-control equipment.

Our Wedding

Mother was terribly afraid that I was going to be late for my wedding at St. John's Episcopal Church in Beverly Farms on June 13, 1942. "You know, that Beverly bridge is oftentimes open. Supposing you get there and you can't cross the bridge?" She insisted that my brother and I leave our house in Marblehead two hours before the service. When we reached the church, we had to kill an hour and a half. Fortunately the minister, who knew my field was aeronautics, got me going on how a propeller drives an airplane. Otherwise, it would have been a long wait.

My brother Peter was my best man. Donny was one of my ushers. I had prescribed cutaways for my ushers. Mother thought that Donny, fifteen years old and pretty big, ought to go to the wedding in an Eaton collar and shorts. He announced that under no circumstances would he appear at the wedding dressed in that fashion! He was either going to be an usher in a cutaway, or he wasn't going. My other ushers were Keith Merrill, Caleb Loring, my cousin Jim Seamans, my old Tower School classmate (and the future author of such plays as *Billy Budd*) Louis Coxe, and my two bicycling companions, David Ives and Johnny Brooks.

The church held just so many people. The decisions about who would sit in which pew were hotly debated. About ten members of the Avalon staff were jammed into the back row. At the very last minute, to everyone's surprise and delight, Gene's godfather, Admiral Emory Land, unexpectedly announced that he and his wife, Betty, could make it. That threw off the whole seating arrangement until it was decided to put an additional row of chairs in front of the first pew.

The church was jammed on one of the hottest days on record. Two ministers performed the ceremony—the Reverend Frederick Morris from our church in Salem and the minister at St. John's, the Reverend Bradford Burnham. I remember seeing Jim Seamans in his Navy uniform, looking white as a sheet. He was apparently still recovering from the lively bachelor's party several nights before. I thought he was going to faint. About the only other thing I remember of the service was feeling very, very pleased as Gene and I left the church and walked outside.

The reception at Avalon was a great affair and a wonderful family gathering. There were two rooms on the second floor filled with beautiful wedding gifts. Mrs. Merrill's elderly friend Mrs. Linthicum from Baltimore was covered with more jewels than I had ever seen. She asked Gene to select any of her jewels as a wedding gift. Gene found that somewhat embarrassing but finally chose a diamond-and-emerald lizard. Mrs. Linthicum promised to send it, but nothing was ever delivered!

I had arranged for Gene and me to stay at the Hart House, an old country inn outside Ipswich. After driving there together in a getaway car we had carefully hidden from pranksters, I set about doing what I felt I had to do: carry my bride across the threshold. But the front doorway of the Hart House was very narrow, and the granite steps leading up to it were steep. Mr. Eades, the manager, held the door open for me, but I failed to maneuver Gene and myself through the door. So he invited us to use the back door. "That will be just fine," I said. What he didn't tell me was that the back door opened directly into a screened porch filled with diners, who were surprised and delighted to watch a bride and groom perform for them.

We stayed at the Hart House for a couple of days, then went directly to the North Station in Boston, where we boarded a train for Lodge Grass, Montana. We spent a week with Nan Bosson Duell, my first cousin, and her husband, Wally, at their ranch on the Crow reservation. Then we headed for Pasadena. We spent a night in Butte, Montana, waiting for our train connection to California. We were on the platform in a small crowd of people at six o'clock the following morning when a man staggered over to us and said loudly, "I think we've got some newlyweds here! How long have you been married?" We had been married exactly one week.

We were met at the Los Angeles station at 7 a.m. by about thirty of Gene's cousins, curious to meet the groom and brimming with warm enthusiasm. That summer they included us in many happy events. After purchasing a temperamental 1936 Dodge coupe, we were able to explore that area of California. We spent several days in Indio with Gene's aunt, Bea Patton. In the 110-degree heat, she provided a retreat for her husband, the general, who had established the Desert Training Center nearby in order to condition thousands of troops for the impending North African invasion. We rode in a tank, and Gene heard the general—in true blood-and-guts style—make an awesome speech to the troops under a broiling sun.

War Work

For the duration of our stay in southern California, some of Gene's elder relatives loaned us their beautiful Spanish villa in Pasadena. Its courtyard was filled with blooming gardenias. While Gene got us settled in our new surroundings, I took on an interesting job with United Geophysical, a company owned by Herbert Hoover, Jr., son of the former President. United had bought from Sperry Corporation the rights to the vibration pickups I had been working on in the Instrumentation Laboratory. The company had been using equipment similar to Sperry's to explore for oil; now it was getting into flight instrumentation by forming a new company (eventually spun off) called Consolidated Engineering. I spent my summer helping Consolidated get established.

When Gene and I returned from the west coast in September, I worked for about a year with Walter McKay, a professor in instrumentation. Our project was to help determine why some Navy aircraft were falling apart or going out of control when they went into or pulled out of high-speed dives. We worked on developing instruments to be mounted in a test plane. We carefully calibrated the air-speed meter and the altimeter, and we developed a couple of new instruments to measure angular acceleration. Obviously, we couldn't ask a test pilot to go up and "just keep her going, though, of course, she might fall apart." So we planned to fly the test plane by remote control, a novel thing itself in those days. There was no telemetry for sending data back by microwave from an unmanned test plane; so we mounted a high-speed movie camera in the cockpit, then hoped to recover the plane and the film after the flight to determine what was going on. If the experiment failed, at least we wouldn't have killed anybody.

After twelve months of work, the instruments were finally built and put aboard an F-6F at Cherry Point, North Carolina. Under remote control, the aircraft staggered up to an altitude of 30,000 feet and started its dive. All the electronics went immediately haywire. The only observation we got out of the whole effort was that when the plane hit the water, it was going very fast.

The experiment was never repeated, largely because Walter and I were already at work on a new project, the A-1 bomb gunsight. We worked under contract with the armament laboratory at Wright Field in Dayton, Ohio. The gunsight was for use in air-to-air combat and air-to-ground strafing. Its purpose was to provide optical information to a fighter pilot so that he could properly lead his target and shoot it down.

Our project officer was Colonel Leighton "Lee" Davis, a key figure in the history of the Instrumentation Laboratory. Lee wanted to move the program fast. In a matter of four or five months, we had put together an experimental gunsight and its electronics. Once the military started installing the equipment in a P-38 and an A-26, we tested it in various locations, including Eglin Field in Florida. To get there, we took Lee's C-45, a twin-engine Beechcraft. He used to let me fly the plane while he went back and worked in the cabin. I did not have a pilot's license; nevertheless, he taught me a bit of maneuvering so that I could go around clouds to avoid passing through them without radar. Once he came out of the cabin and said, "Maybe you don't need to be quite so strenuous with the controls!" On another occasion, I was flying along when suddenly all of the bells and whistles in the cockpit went off at once. Lee came tearing out of the cabin and jumped into the pilot's seat. One fuel tank had run dry and the engines had shut down. Fortunately he was able to transfer to a second tank and restart the engines.

When we arrived at our destination and the equipment was tried out by test pilots, I often went along for the ride. Some pilots, just back from the Pacific, loved to fly at an altitude of about twenty-five feet. I watched with alarm to see whether the tips of our propeller were going to hit the ground, as we made our dives and passes, firing five-inch rockets.

Lee Davis's career and mine ran roughly in parallel. Later, when I was working on the Tracking Control Project, he was the chief of the armament laboratory at Wright Field, where our work was put to the test. Later still, he was the Air Force's number-one man at Cape Canaveral when I was working in the space program at NASA.

While working on these various wartime contracts, I was also teaching in the Navy's V-7 program at MIT, the same program I had wanted to enter. Officers-in-training were with us for three months, hence their nickname "ninety-day wonders." The instrumentation class, which I taught, was five days a week for six weeks. I taught that course for thirteen consecutive six-week sessions. I would finish one section on a Friday and start the next group the following Monday.

The first time I taught the course, I was probably the youngest person in the room. There were some smart alecks from industry in the back row, who tried to ask me impossible questions, but I learned very early how to deal with them. I would announce to the class that we were very fortunate to have Mr. So-and-so from the XYZ Corporation with us. I then invited him to give one of the lectures. I found that shut him up quickly.

Target Acquisition

In late 1944 the Navy called Doc Draper with a so-called target-acquisition problem. Japanese kamikazes were making it especially difficult for our Navy gunners by attacking with the Sun behind them. A gunner couldn't look up into the blinding Pacific sun and "acquire his target" (see a kamikaze long enough to get it lined up in his gunsight). Radar operated out of the CIC (command-in-control room) could pick up any incoming plane or missile, but how to translate this radar information to the gunner stationed at a gun mount next to the flight deck? I went with Doc Draper down to Sperry Corporation on Long Island and watched in awe as Doc stood before a blackboard and developed extemporaneously a system to solve the problem.

It was my job to flesh out the design for this equipment, get it built, draw a wiring diagram, and be down at the Navy yard in Bayonne, New Jersey, to go aboard the *Bonhomme Richard* in three weeks, starting from scratch! Just getting to Bayonne, New Jersey, loaded with equipment was hard enough in those days, because wartime travel restrictions made it almost impossible to fly anywhere, and taking the train involved connections, subways, taxis—while lugging the equipment all the time. I finally got myself a Hertz rental car, packed up, and set off.

When I reached Bayonne, no one would let me in the Navy yard at first. I finally talked my way in and went aboard the ship. After explaining my mission, I was steered to the gunnery officer. When I finally got to see him, he said under no circumstances was anybody going to touch anything on that ship! There were green tickets on everything, meaning that the ship had been fully checked out and was ready to sail. He didn't want me messing with anything.

I got to a phone booth and called Doc Draper. The next day Admiral Martel from the Bureau of Ordnance arrived at the Navy yard. I was told to bring the equipment aboard. Before I could get it installed, we set sail. Within hours, I found myself working on an open deck in the middle of the Atlantic Ocean in the middle of winter. I had only one copy of my wiring diagram, drawn on a huge piece of tracing paper. If it had ever blown overboard, I would have been in deep trouble. To tie the system in with the ship's compass, I patched into a junction box just over the captain's bunk and learned he was upset. When tying into the ship's radar, I opened up other junction boxes and made my connections. I came back an hour later to find that the technician from Western Electric, the company that had installed the radar, had disconnected all my wires.

About three days later in the Caribbean, Admiral Martel and some other brass came out to see how the equipment operated. Everybody seemed to be pretty pleased, and I was cleared to go ashore. The ship anchored off Norfolk, Virginia, and I clambered down a rope ladder on the side of the ship and into a small boat, clutching my oscilloscopes and other test equipment. When I finally got to a phone to call Doc Draper and tell him what had happened, his first words were: "Damn it, can you tell me where that Hertz car is? We're paying through the nose and they want it back!" The car, of course, was still parked at the Navy Yard in Bayonne, New Jersey, where I had boarded the *Bonhomme Richard*.

I traveled back to MIT by way of the Bureau of Ordnance, where I delivered a copy of my wiring diagram. On my return to Cambridge, Doc said, "There's a destroyer about to set sail from the Charlestown Navy Yard. They're having some trouble with the radar. Get over there and help them out." So I worked on the equipment in Charlestown for a couple of days, then went back out to sea aboard the SS *Purdy*. After several very cold days, I was finally ashore and the *Purdy* was headed for the Pacific, where it shot down a good many kamikazes before being sunk.

Tracking Control

After the war, I became involved with another project sponsored by Lee Davis. In preliminary discussions, he told Doc Draper and myself that he wanted to develop an improved sight to be tied in with tracking radar, which was just coming into its own. The question was, once a pilot had acquired a target and locked on, under what conditions could a plane be made to maneuver automatically to track the target until the target had been fired on and destroyed? To solve this problem it was clearly necessary to study the dynamics of the pursuing airplane itself. Did existing planes have enough maneuverability to follow the evasive action of the pursued plane?

Every project at MIT had to be cleared through Nat Sage, the director of research. He, Doc, and I discussed it around the table, and Nat Sage finally said, "Put it in a letter to me." That night I wrote a two-page letter. The following day I showed it to Doc and took it over to Nat. That letter became the basis for the Air Force contract I headed for the next four years. It became known as the Tracking Control Project. It was the largest project I had headed to date. I had about thirty-five or forty full-time people working on it, as well as nearly one hundred master's and doctoral candidates assisting in the research.

There were many parts to the problem. For one, we had to study how quickly the plane reacted to movements of its stick. For another, we had to run the same kinds of tests on the servo-activators that transformed electrical signals into movements of the elevator, rudder, and ailerons. Then we did the same with radar, to see how well existing equipment could follow a target. We built our own simulator (electronic analog computer) and some of the flight hardware out of whole cloth. A number of disciplines were involved in creating and testing the system. Simultaneously with this project, I was starting a graduate program in the automatic control of aircraft. It was a lot of work and a lot of fun.

Eventually we had a complete system, which, when installed in a plane, could lock onto and follow an acquired target. Three or four hundred people were invited to Wright Field for the first large-scale demonstration. They were invited first into an amphitheater where I, together with others who worked with me, explained the system. Then everybody went outside and looked up as the target plane came overhead and our tracking plane locked on and followed. I hardly dared look, because we had been under considerable pressure to rush the job and hadn't tested it many times prior to this demonstration. It worked, though it seemed to me, as I was watching, that the pursuing plane broke lock momentarily and that the pilot took over manually until the system relocked. I never dared ask Chip Collins, our test pilot, if that is what had happened.

As a result of our team's work on tracking control—for developing what was termed a "dynamical model" of an airplane—I received the Elmer Sperry Award, given in memory of a Sperry family member to a young engineer for outstanding contribution to the field of aeronautics. I used the stipend that came with the award to buy pewter mugs and to have them inscribed for the key people who had worked with me on the project.

Middleton

Middleton in the 1940s was nearly open land located between Boston and Andover. For the Seamans family, Middleton provided an opportunity for country living as we moved from Cambridge in 1948 and, three years later, to our permanent future home on the seacoast in Beverly Farms. Kathy, Toby, and Joe were born during our days in Cambridge; May and Dan arrived after we settled in Beverly Farms.

In the spring of 1945, Gene was six months' pregnant and fell while carrying Kathy downstairs. Two days later she began to have contractions, and I rushed her to the hospital. The doctor assured us she would be all right. "But she's going to lose the baby," he added. On questioning he advised us there wasn't one chance in 10,000 the baby could survive. About two o'clock in the morning on May 22, 1945, a whole group of us were sitting in the waiting room at the Phillips House at Massachusetts General Hospital. Dr. Titus trundled in and said, "Mrs. Seamans is just fine."

After a silence, Gene's mother asked, "And the baby?"

"A very well-formed little boy," Dr. Titus answered. But Toby was three months premature, almost insurmountable odds in those days. We had already tipped off our pediatrician, Dr. Joseph Garland, and he had been boning up on what to do with extremely premature infants. Within a couple of hours, he was attending our son. Toby weighed three pounds, three ounces, and was twelve inches long. He had no eyebrows, no fingernails, and no toenails, and his head was about the size of a small orange. The nurses had to rotate him regularly, because his head was so soft that it would flatten if left in one position. They had to be terribly clever to feed him. A tiny rubber tube attached to a test tube of milk was moistened and inserted straight down his esophagus into his stomach—carefully, of course, so that it did not enter his lungs instead. The milk was then poured in, and the tube was pulled out. The whole operation happened in a matter of seconds.

I asked Dr. Garland for his prognosis. "I think you've got one chance in ten. It depends on a couple of things," he added. "Obviously a lot of it is physical—how strong his constitution is. But even if he's the strongest baby in the world, an upset stomach could tip the scales, because he has so little margin. Still, what really matters is his will to live." It was hard for me to believe that this tiny being in a bassinet could have a will to live, but Dr. Garland evidently believed it, and now I do as well. Two months later Toby left the hospital weighing just over five pounds. At age three he was a healthy, cheery child, but he had no speech. We were told by a specialist that he'd never seen a deafer child and he would never talk.

A long crusade followed with Gene in the lead. Thanks to Helen Patten and the Winthrop Foundation of the Massachusetts Eye and Ear Clinic, his speech is remarkably good. He has a bachelor's degree from Columbia and an MBA from Northeastern.

Toby and Stella Mae Renchard were married on December 27, 1969, before his graduation from Columbia. The small wedding was held in the Chapel of Apostolic Delegation in Washington, D.C., with Helen Patten and immediate family in attendance. The Papal Nuncio officiated.

Afterwards at the home of the bride, we were welcomed to a spiffy dinner by her parents, George and Stella Renchard. With a long, distinguished career in the State Department, Ambassador Renchard had known Gene's father early in the 1930s. He had just ended his last tour of duty as U.S. ambassador to Burundi.

Toby's situation, as it developed through the late 1940s and 1950s, was extremely important to Gene and myself, and it influenced various career decisions I was faced with along the way. There were jobs I turned down (happily, in retrospect) because we didn't want to slow

Toby's progress by taking him away from Helen Patten. His success has been an inspiration for the whole family. Toby is a most perceptive stock analyst. He and Stella Mae have five children, and they live near us in Beverly Farms.

We moved to Middleton shortly after Joe was born. We had been forever trying to get to the country on weekends, and we finally said, "Let's be there all week!" We looked casually at places in Beverly and Marblehead before my distant uncle, John Pickering, told us we could rent his wonderful old farmhouse.

None of our friends could understand why we would want to live that far out in the country. Conditions were a bit rustic. We had to rely on an old stone well for water. Once, we had invited a number of our friends out to the house after a Harvard-Princeton game. Two days before the game the well had gone dry. The only way to get water was to bring it back in buckets from a hydrant about a mile away. All during the evening, men could be seen heading outside and behind the trees.

We lived in the country at a perfect time, when our older children were young and could enjoy the fifty acres at our disposal. There was a pond across the street where we enjoyed wonderful skating in the wintertime. Our daughter Kathy, who was very independent, became an outdoor enthusiast and prolific reader. One day, in a fit of exasperation, she took off on her tricycle, unbeknownst to us, and headed down the country road. More than halfway to Middleton center, she met a policeman who asked where she was going. "To town," she responded. She was brought home crestfallen. Toby, still struggling with speech, cheerfully enjoyed our livestock of chickens, ducks, turkeys, and geese. He and his sister were delighted to discover that at the soda fountain in Middleton, customers had stuck their chewing gum under the counter. At ages three and five, respectively, they saw it as a free source of candy! Meanwhile Joe, still a toddler, developed his ability to keep up and fit in.

Gene loved the country, while barely catching her breath the whole time. For more than five years, summers were spent at Avalon, the Merrills' lovely place at Prides Crossing. All the grandchildren convened there to enjoy seaside activities, each other, and their grandparents, who did an amazing job of running a household of over twenty people and providing wonderful meals from the large vegetable garden.

It was in Middleton that I did most of my doctoral work. Dr. Jerome Hunsaker, chairman of MIT's aeronautics department, came to me in 1948 and said, "Look, you're going up this track pretty fast. Pretty soon it will be unreasonable for you to go after your doctor's degree. So if you're really interested in getting a doctorate, you had better get going."

Gene and I had quite a discussion about this because we realized what it would mean to our family life. My friend John Sluder was instrumental in my final decision to go ahead with the doctoral program. His wife, Betsy (Bradley), was a great friend of Romey's. John, who had his Ph.D. from MIT, was eleven years older than I and was a great role model for me.

I told John what Hunsaker had said. His response was, "You've got to go for it."

"Why?"

"You think you're always going to want to stay at MIT. If that were true, you wouldn't need it. You're already in the chain of promotion, and you'll work your way up, no doubt about it. But what if you want to leave MIT some day?"

Gene and I agreed that I shouldn't take off three years from work to concentrate on my studies, but that I should continue teaching (the course on automatic control of aircraft for about forty graduate students) and managing the Tracking Control Project. On top of which, I was advisor on about two dozen master's theses during the three-year period. Furthermore, I was just starting to get involved in a little outside work with the National Advisory Committee for Aeronautics (NACA). In short, I was quite busy.

My field was instrumentation, a multidisciplinary study program overseen by the heads of the departments of physics, math, electrical engineering, and aeronautics (of which Doc Draper was about to become the head). I had to satisfy the requirements of each department. I also had to take a minor, for which I chose mathematics. Finally, I had to study two foreign languages. I had the French, but I had to start from scratch with German.

So there I was on a typical Saturday at our old farmhouse, wrestling with equations, when our three kids came rushing in asking me to go out and play. Although I wanted to play with them, it often took me a half-hour to get my train of thought back to where it had been before the interruption. Finally, Gene made a big sign for our bedroom door. It read: "Daddy is working. Please be quiet." From then on I often sat at my desk in our room and heard the kids running up and down outside—Kathy, the eldest, followed by Toby pulling Joe. They would arrive at the door, and Kathy, in a loud, very high-pitched voice, would say, "Daddy is working. Please be quiet!" That only took me about five minutes to recover from. When I finally got my degree in 1951, we made a big ceremony of taking that sign out and burning it.

My oral exam, which followed the written ones, was almost a disaster. I had five examiners, the heads of the five departments overseeing my doctoral work. The head of the physics department was a meticulous professor named Sears, who expected precise answers. I had had an acknowledged deficiency in physics when I first came to MIT to do graduate work. The physics department had felt that the Harvard physics program wasn't up to snuff, and I had taken MIT's freshman course. I was able to score 100s in all the exams and successfully petitioned the department not to have to take the sophomore course.

Now here I was in the oral exam; Professor Sears was asking me questions, and that sophomore course would have helped! One question I was completely unprepared for was: "Name five or six physical constants and indicate how they can be measured." The only one I could think of was the gravitational constant, and nobody had ever told me how it's measured. I guessed and got it right. Sears said, "All you engineers come up with that one first, but now I would like to have you name five more." I could not name one. I thought, "Boy, that's it! It's curtains for Seamans and his doctorate!" Fortunately, I did well enough on balance with the other faculty members present that I was allowed to pass.

My thesis, which was then classified, was a further development of the work we had started with the Tracking Control Project. It compared several methods for an interceptor airplane to track a target.

My Mentor

Of all the people who have had an influence on the way I've thought, apart from my family, Doc Draper is preeminent. From 1940 to 1950

I worked for him, first as a student, then as a colleague in the Instrumentation Laboratory (which was subsequently renamed the Charles Stark Draper Laboratory in his honor and was divested from MIT in 1973). Doc Draper was not only tops in the field of instrumentation, he was a remarkable role model. Doc was not a person who liked sitting at a desk all day. He used to love to run his laboratory, working night and day himself. I've never seen such a prodigious worker. He would take people out to dinner at a place called the Fox and Hounds, have a reasonable number of martinis and a dinner, then come back to the office and continue working until two or three in the morning. First thing in the morning, he was on the job when everybody else arrived for work. I don't know how he had the energy to do it. He worked Saturdays and Sundays throughout his whole career.

He was extremely practical. His approach was always to figure out where the critical elements of a problem were and then to place his and the laboratory's emphasis there. He deeply believed that theory and practice must always be kept in balance. You can't proceed only by trial and error, nor can you get results by merely theorizing. It's a constant back-and-forth between the two that solves problems.

Doc had plenty of ingenuity. When he was working on his first gunsight, the Mark 14, the Navy told Doc that he had to demonstrate that it was producible. He tried to get various companies to manufacture it but couldn't find anybody to take it on. Finally he located a machine shop in Newton owned by Fred MacLoed and John Sattlemeyer, and gave them the task of building fifty complete gyroscopes. The machine shop didn't have a name, so Doc said, "We'll call it Doelcam," which is almost MacLoed spelled backwards.

John Sattlemeyer was a straight, thoughtful man and a very good machinist. Fred MacLoed was expansive and more the entrepreneur. They made their fifty gyroscopes and got a big contract out of it. The two men eventually became multimillionaires. Sattlemeyer never changed his straight and simple lifestyle. Fred MacLoed left a wife and seven or eight children for other women and race cars.

Doc Draper went through periods when he was very, very difficult to work with—whenever he was working on a problem in the back of his mind. At these times, he would become very impatient with the day-to-day administration he had to oversee. Nat Sage would say,

"I can always tell when Doc's going to hatch a new one. He becomes ornery!" Yet, when it came to management, Doc did the things that were absolutely essential, and he did them right. Even when he was embroiled in a problem, he showed great day-to-day concern for the people who worked with him. All of them. It didn't matter who they were. If somebody had a key job on a milling machine, Doc would learn his first name and visit with him at his bench. He held open houses at his home and invited hundreds of staff and students at a time, from all levels of the laboratory and department.

There are so many wonderful stories about Doc. Whenever he explained a project to a group of laymen (military procurement people, executives from industry, and so on), he did so in simple terms, but he always threw in at least one very complicated chart. This way, even though people understood most of what he was talking about, they never knew everything. He wanted them to realize it wasn't easy! He did this quite intentionally. Once when he was up in front of a group explaining a target-acquisition problem, he drew a very complicated diagram of lines and boxes representing servos and computer. Suddenly I wasn't sure that even he himself knew exactly how to build the system!

Doc and I once spent a month together at a naval anti-aircraft test facility near Virginia Beach. We stayed at a place on the beach called the Gay Manor. We called it the Gay Manure, a better description of the food, we thought. We slept in twin beds in the same room. At night, Doc liked to read paperback science fiction. "One thing you never want to do," he would say, looking over at me, "is bring along a very good book on a job like this, because you'll tend to read it and stay awake too long, then be tired the next day. You need a book that's so bad you'll fall asleep on the third page." Sure enough, he started reading, and before long he had conked out with the book on his chest. The next morning we would be up at six o'clock, have a quick breakfast, tear out to the firing range, work there all day, be back in the evening, and have drinks and dinner with our associates.

Sometimes when visitors came to the laboratory, Doc would say, "You know, we make it a rule here that we've got to keep things moving along until 5:30. No cocktails or anything until then." What the visitors didn't realize was that Doc had underneath his desk a button with which he could control the speed of the clock. So if they arrived at, say, 4:30, they might find that after fifteen minutes it was 5:30 by Doc's clock. "Marie!," Doc would yell to his secretary. "Come in and take the drink order." He loved to play this game.

As with any mentor, I learned innumerable things from Doc Draper. But there were things I could never learn until I had left the shelter of his wing, which I did in 1950. It was only then that I began to learn how to coordinate the work of large numbers of people grouped in different disciplines. I had to learn the hard way by making mistakes, some of them bad mistakes. Fortunately I got this experience at MIT, because such an understanding would be absolutely essential to success at NASA.

Project Meteor

In 1950, I was approached by Doc about becoming systems engineer on Project Meteor, an effort to assimilate Germany's experience with the V-2 rocket and build rockets of our own. It was known throughout the world that the United States was ahead in many areas of technology. But we were behind in missiles, compared to what the Germans had accomplished during World War II. Project Meteor had been initiated at MIT by the Navy's Bureau of Ordnance right after the war. It had initially been viewed as an effort to look at the new technologies in missilery, thus far unknown to the United States.

Six or seven departments at MIT got involved, with each working on a particular set of technologies. The servo lab people designed and built new hydraulic, high-performance actuators, and the propulsion people looked into new kinds of rocket propellants. A wind tunnel was built requiring so much energy, it couldn't be turned on without calling the electric company and asking when they would have enough power to run it. To keep it cool, so much water had to be pumped in and out that there was reason to worry about erosion of the banks of the Charles River.

After three or four years, the Navy turned around and said, "This is great work you're doing, but where's the missile?" It was with a view to pulling together all these strands of work that I was installed as systems engineer. Why me? In part, I think, because of what I had been doing on the Tracking Control Project—another case of trying to build a total system out of disparate parts. Project Meteor was a difficult task. I recognized that my hold over the contributors from each of the departments at MIT was a lot more ephemeral than what I had grown used to while working in the Instrumentation Lab, and I didn't always handle this situation with great finesse.

I decided that I had to have a central systems group that could spell out what each department was supposed to be working on. I needed specific expertise in this central group that understood what all the individual groups were doing. I had in mind transferring several people from their respective departments. Some of the departments were reluctant to give up staff, and I got pretty upset. I thought that if I was going to take on this responsibility on behalf of the institute, people ought to be more forthcoming. I had a meeting with the heads of all involved departments in president James Killian's office and mentioned the name of someone I wanted transferred. Someone asked, "Well, supposing he doesn't want to come and work for you?"

I said, "I think it's time we got a little discipline in this place." The whole meeting exploded. Clearly, that had been the wrong answer. I was violating the principle of academic freedom! I should have said, "Would you mind my exploring the possibility with him?" We finally managed to extract some people from the various departments and hired some others, putting together a nucleus of about seventy-five people, which I felt we needed to develop a missile.

In time, the Eisenhower administration became concerned that America was falling behind the Russians in missilery. K. T. Keller, who had run Chrysler, was made "missile czar." He came up to MIT to assess the status of the Meteor missile and was joined there by people from the Navy, as well as associated contractors like Bell Aircraft. By then we had built some experimental dummies and had fired them, but they weren't missiles. We didn't have a missile. The result of the meeting was that the Navy decided Bell Aircraft would begin building an air-to-air missile that didn't yet exist!

The situation became complicated. Bell Aircraft was eager to mass-produce missiles. The Vitro Corporation was brought in to test the missiles and other equipment and to train Navy personnel in their use. Meanwhile, we were in the middle, a bunch of research laboratories at MIT technically responsible for the outcome of the program. The whole project grew to a tremendous size before we even knew what we wanted to build. I had some serious arguments at Bell Aircraft involving Larry Bell and naval personnel. I felt that the project had quickly gotten out of hand, and I got a reputation as something of an enfant terrible for trying to rein it in. Unfortunately, Doc Draper was not directly involved, and the program lacked a real godfather. It had focus in each department, but there was no central authority figure, like a Doc Draper, to get the Navy and the contractors to cease and desist until we had a prototype design.

In the end, the design people at Bell built what I called an "inside-the-egg missile." You couldn't get at any part of the missile without moving five other parts. You couldn't take out the servo package or the gyro package and work on that alone.

The first test of our air-to-air missile took place on the west coast. The plan was to fire the missile from beneath the wing of an airplane. Afterwards, I got a call from the man in charge of the testing. "It's gone," he said.

"What do you mean, it's gone?"

"The missile's gone." As the plane carrying the missile came in towards the shore, our engineer, who was sitting in the plane, looked out and saw that the missile was gone. It had been released inadvertently. The Navy sent groups of seamen walking over bean fields where the missile was thought to have come down. They stumbled onto a huge hole in the ground. The area was classified, and some heavy excavation equipment was brought out. Fifteen feet down, there was our Meteor missile—and not in very good shape.

Subsequent tests were successful, but it was not long after this that I got a call from Nat Sage, who said, "We've been torpedoed in the engine room. Project Meteor is canceled." The Bureau of Ordnance had given us three months to stop all work. The next day Nat and I went down to Washington to see what we could salvage. All we could get was $50,000 that had already been budgeted for Meteor, in order to write a final report.

On its face, Project Meteor was a failure. We did not provide a missile for the Navy. We did launch the Meteor from beneath the wing of an airplane a number of times, but by the time of cancellation the missile still had not homed on a target. When we came to the very last day of the project, still without success, I gave orders to the people at the test field to close up shop. The next day I got a call from the group leader, Bob Briggs. He was ecstatic. "It worked! It worked!," he shouted. This successful test failed to save Project Meteor, but at least it provided some psychic rewards for the people involved, and it made our final report both more positive and more useful to ongoing missile research.

Our efforts were not completely wasted. The Applied Physics Lab at Johns Hopkins, which continued working on ship-to-air missiles, was able to put some of our thinking to good use. Furthermore, we approached the Air Force about picking up our end of the project, and though they didn't do so, they used bits and pieces of our technology. The anti-radiation missiles in use by the Air Force today employ some of the system thinking we were working on at MIT in the early 1950s.

RCA

At MIT, the groups that had been getting money from the Navy under Project Meteor had to go their own ways, each figuring out whether it was going to retrench or not. My systems-engineering group had come together for the sole purpose of working on Meteor and had no departmental affiliation. So I began looking to see what other kinds of projects might be appropriate for this group. In pretty short order, we picked up three new contracts. With these in hand, we called ourselves the Flight Control Laboratory.

We proceeded with our contracts for about eight months, or until the summer of 1954. The administration of MIT withheld full approval for the Flight Control Laboratory on a permanent basis, and when Gene and I left for a summer cruise off the coast of Maine, I knew that Nat Sage and Doc Draper were talking with James R. Killian about the future of the Flight Control Lab. One night Donny, Bev, Gene, and I were anchored in Pulpit Harbor in pouring rain when somebody came alongside and said, "Is Seamans aboard?" It was Dave Wheatland, a friend of my parents. He told me that Doc Draper was trying to get me on the phone, then took me to his farmhouse on North Haven, where I placed a fateful call.

"We did the best we could," Doc told me, "but it's all over. You've got to close down the Flight Control Laboratory." We were given until the following June to wrap things up, so that we could complete the various studies for which we had contracted. While this situation was still fluid, I had a call from John Woodward at RCA, who explained that his company had an engineering development program in Camden, New Jersey. Would I be interested in helping them run it? I had been asked to consider leaving MIT before, but none of the opportunities offered had been acceptable to Gene and me. The RCA situation was close enough to my expertise that I couldn't reject the offer outright, so I said to Woodward, "I can't answer that on the phone."

"Then come on down to Camden and see what's involved," he responded. We set an appointment.

Camden was a typical industrial city of that era. Approaching it from Philadelphia, which I knew as an attractive city, I crossed a bridge and felt as though I were plunging back into the Industrial Revolution. Camden had a lot of brick buildings, including quite a few owned by RCA. Campbell's Soup was located there as well. The whole city seemed to smell of soup.

I visited RCA's corporate headquarters. John Woodward introduced me to Ted Smith, the executive vice president in charge of defense products. I was offered a salary ($15,000) that was more than I was making at MIT ($10,000)—not a bad offer, but not overwhelming either. I took a look at the surroundings and couldn't see the Seamans family living there. Nor could I see myself immersed in what was clearly a large organization and trying to mold something within that environment. Whenever a decision like this came along, I never said yea or nay until I returned home and talked with Gene. This was not a difficult decision for the two of us. I called Ted Smith a day or two later and said, "No, thank you."

About a week later a very excited John Woodward called me on the phone and said, "We've been thinking this over. We know that you've got people in your laboratory that you've been working with. What would you think of starting an RCA laboratory in the Boston area with some of these people?" That was a different proposition in many, many ways. It satisfied several of my concerns, including what to do with the people working with me, not to mention what to do with myself! I couldn't see full-time teaching, in part because I continued to believe that teaching, lab work, and thesis work were all of the same cloth and enhanced each other. The long and short of it was, I took the job after negotiating a salary of $18,000.

With a lot of hoopla, RCA kicked off their new laboratory in December 1954 with a big reception at the Hotel Statler in Boston.

Up from RCA New York for the occasion was Elmer Engstrom, a fine person in charge of engineering throughout the corporation. The laboratory was introduced with much fanfare—with a presentation of the imaginative new program the company was embarking on and an introduction of the young man who was willing to step in and take charge of it. (I didn't think of myself as a young man at all.) They even brought in a robot to liven things up. It came rolling in through the door, went up to the podium, and handed a message to Engstrom.

The next six months were hectic. I oversaw the start-up of RCA's Airborne Systems Laboratory, while also phasing out the Flight Control Laboratory and teaching a full load at MIT. First we had to figure out where to locate the lab. It was decided to take over a big section of the old Waltham Watch Company building, which was in pretty rough shape when we took occupancy. The lavatories were the crummiest I had ever seen, with signs on the doors reading: "Your Management Takes Pride in the Facilities It Provides Its Employees." In time, RCA did a nice job rehabilitating the building.

Then we had to go out and recruit the people to fill the lab. RCA was doubtful that we could fill all the slots. We needed to find thirty people a month for three months to get started. I said I was sure we could do so. At MIT, I had worked extensively with Joseph Aronson, who had been the U.S. Air Force Air Materiel Command representative there and then had been an assistant director in the Instrumentation Laboratory. He was one of the first people I hired, and he oversaw much of the recruiting while I was still completing my assignments at MIT. About half of the employees of the Flight Control Laboratory ended up coming along with us. We recruited locally for most of the rest.

Once things were moving in the laboratory, the company threw another big reception. We had an open house and a dinner, at which I was introduced. State government officials were present. Dick Preston was on hand, as secretary of commerce and a representative of the governor. He had great wit and made a hit when he spoke. After dinner, big color-TV sets were brought in, and we all watched a color broadcast of *Peter Pan* starring Mary Martin. The RCA system for color TV had recently been accepted over that of CBS, and the company was actively promoting its system.

Airborne Systems Laboratory

In the early 1950s, the Hughes Aircraft Company had a monopoly on fighter-aircraft fire control—the whole system for aiming a fighter's guns and rockets. As the only game in town, Hughes could hold the Air Force up for ransom. Well before I arrived on the scene, RCA had received a contract to serve as a second production source for Hughes equipment. That contract allowed RCA, if the company wished, to use some of the overhead costs to improve on Hughes's system. By the time John Woodward and Ted Smith proposed that I set up a new lab, the Air Force had been putting pressure on RCA to come up with something more imaginative, a more advanced technology. That became the primary mission of our lab. Our contract number was 28007. That became our identity. Whenever someone at RCA asked what we were doing, the answer was: "We're working on twenty-eight double-oh seven."

Eventually we got some other contracts, and we began outgrowing the Waltham Watch building, which wasn't an ideal place anyway. RCA agreed to put up a new building for us. I started dealing with people from headquarters about buying land. The company still had a requirement that any land purchased have a railroad siding adjacent to it. The property that looked most attractive—on the Middlesex Turnpike (Route 3A) off Route 128—had no railroad spur. We finally got the company to agree that we didn't need one. RCA bought thirty-five acres in Burlington, Massachusetts, and I helped get the land rezoned for industrial use.

Then we had to design the building. I had definite ideas on the kind of building I wanted—three or four stories tall and as compact as possible, so that people could get from one end to the other quickly. I wanted everyone to have a window to look out, to take advantage of the country landscape we were investing in. So I recommended a four-story, H-shaped building.

The first design that crossed my desk was for one story. "Why one story?," I asked.

"We're still not sure you're going to succeed, and if you don't succeed we want to be able to convert the building to a warehouse."

I continued to press for an H-shaped building and came up with what seemed like a great design. To my horror, when I arrived at the building meeting in Camden and looked at the design that had been worked up, its floor plan was square.

"But we talked about an H!" I said.

"It really is an H," the designer said. "We just sort of folded it together."

I asked Johnny Woodward if we could go and see Art Malcarney (who by then had replaced Ted Smith as head of defense products). When we did, Art said, "If Seamans feels an H is what's needed to attract the people that we want, make it an H-shaped building!" We got a one-story H.

Then, right in the middle of construction, contract 28007 was canceled. We had reached the point where to continue with what we were doing would have involved a major increase in funding for RCA, and the feeling at the Air Force was that we already had accomplished our mission, namely, to put pressure on Hughes. Now that the Air Force had Hughes where it wanted them, it was time for us to get off the stage. It was not that we had failed the test. We had come up with designs that could have been used.

We had other contracts by then, but the cancellation of 28007 meant we had to make a big reduction in staff. We were still in Waltham at the time and were roughly six months away from moving to Burlington. I called a noontime meeting of the entire lab staff at Nuttings on the Charles, a place just around the corner from the Waltham Watch building, where I had gone in my college years to see Benny Goodman and other big bands. I intended to announce that, because of the cancellation of 28007, we had to let a third of the people go and that those who were being furloughed would receive their pink slips that afternoon. Just as I was heading to the meeting, I got a call from somebody in Camden who worked for Johnny Woodward. He said, "We think you ought to cut down by 50 percent, because even the contracts that we have are beginning to look a little bit shaky."

I said, "If you want to cut 50 percent, you come up here and do it yourself, and I'll tender my resignation." Headquarters finally backed off, and we let a third go. Still it was pretty traumatic.

A week before our new building was set to open, Malcarney came up to look at it. I had complained about the raw cinderblock walls in the corridors. The place looked like a jail. I had been told that cinderblock was the most efficient material and didn't have to be painted. Paint cost money. On his inspection tour, Malcarney started walking down a corridor, stopped in his tracks, and said, "What damned fool around RCA leaves a wall looking like this?"

"What color would you like to see?," I asked him.

"Yellow."

"Yes, sir."

The dedication ceremony for the new building was quite an affair. Gene and I invited a lot of guests, including my father. RCA had many musicians on its payroll, including Arthur Fiedler, conductor of the Boston Pops. A high school orchestra was assembled for the dedication, and Fiedler conducted it. Governor Foster Furcolo gave a speech and cut the ribbon. He was given a high-tech device with which to zap the ribbon electronically, only it didn't work. Finally a pair of scissors was commandeered so that Furcolo could perform his role. Afterwards we put on a big show to demonstrate the kind of work we were doing.

Now that we had a new building, we had to get more work. RCA did acquire new contracts and eventually built back up to and past the size we had been at before the loss of 28007. Buildings were added to accommodate manufacturing, and by the mid-1960s, the Burlington facility employed about 2,500 people. By then, though, I had moved to a much bigger, much more exciting mission.

Wrapping Up

My five years at RCA, from 1955 to 1960, were my only experience working full-time for industry. The big difference at RCA, compared with MIT, was that everybody in the laboratory knew he or she had one job—to work for RCA—and that I was RCA's manager in residence. There was nobody coming around saying, "I'm sorry, I can't help you today because I'm teaching a course, and tomorrow I'm attending a meeting in Washington." Everybody was pulling in one direction.

On the other hand, a corporation is a maze of people, and at RCA I had to deal with many layers in order to get things done. I had hundreds of people working for and with me. I reported to John Woodward in Camden, who was in charge of aircraft fire control. He in turn worked for Ted Smith (replaced by Art Malcarney), also in

Camden. In addition, there was the New York office, where the whole organization was headed by David Sarnoff, founder of the company. At each level of the organization there were staff people, making the maze that much more complex.

There were a lot of constraints on me as manager. Starting salaries were just one example. When I tried to hire someone for slightly more than the level dictated by RCA's pay scale, I heard from administrative people who said they didn't see how RCA could pay that much.

"He's the perfect person for this job," I said, "and you're going to quibble over $30 a month?"

"I'm sorry. For somebody in that category, the bandwidth goes from $570 to $590." So I had to call someone higher up the chain of command and haggle with him. I generally won, but it was a neverending battle.

I suspect that I was not as well suited to the corporate life as I was to academic life (or later to government work). At RCA, I found it very hard to focus primarily on the bottom line. Obviously, a company can't be in the red year in and year out and stay alive, but at RCA so much seemed to hinge on dollars and cents month by month (and certainly quarter by quarter) that it seemed very difficult to get people to think long-term. If margins were pressured, for example, one of the first things that usually got cut was independent research and development. I thought it ought to be just the reverse. When times get lean, a company ought to spend money on what's going to help it four or five years down the road instead of trying simply to look good next quarter. The securities analysts aren't going to like it, and the stock price is going to go down, but in the long run it may be better for the company.

In Art Malcarney, I found an exceptional role model for management within an organization as big and complex as RCA. Malcarney provided my first experience of dealing with what I would call a hard-boiled businessman. He had come up through manufacturing. He was tough. The people who worked with him stood in awe of him. If he said go, you went.

One story epitomizes him. RCA had the job of designing and building the ground electronics for the Atlas missile. One part of this system was called APCHE (Atlas Programmed Checkout Equipment). APCHE's purpose was to determine whether the Atlas missile was ready to go or not, and manufacture of APCHE became our responsibility in Burlington. The total RCA effort was not going as well as the Air Force wanted, and the Air Force was bearing down on RCA. Instead of telling the head of the division responsible for the Atlas ground electronics to get going, Malcarney personally flew to Vandenburg Air Force Base to oversee RCA's work there. He commandeered a trailer and set up shop. Every morning at six o'clock he held a meeting in his trailer to find out how RCA had done the previous day. If things weren't done, people had to duck!

A piece of equipment was needed urgently. Malcarney called a vice president back in Camden and said, "I want to have that piece of equipment out here tomorrow in the a.m., and I want you to bring it." The vice president got the equipment but thought to himself, Malcarney couldn't mean that I personally am supposed to take it out there! So he arranged to have it flown directly to Los Angeles, picked up there, and driven sixty miles north to Malcarney at Vandenburg. Bad weather came along, and the flight was diverted to Dallas, Texas. Malcarney telephoned the vice president and fired him on the spot. Then he made it very clear throughout all of RCA what he had done.

Tough stuff—but Malcarney saved the project. And while he was out there in that trailer, he was not only running the Atlas electronics project but also overseeing the totality of RCA's effort in defense products. That was quite an example for me. I don't think a manager always has to go to that extreme to lead people effectively, but clearly he has got to tell people what he wants them to do and be prepared to react unequivocally if they don't do it. Otherwise, his leadership will rapidly erode. It was a lesson I have tried to apply everywhere else.

CHAPTER TWO

The NASA Years

ONE week before his assassination in November 1963, President Kennedy said, while flying over Cape Canaveral, "I think the most significant event that took place in the fifties was the launching of Sputnik." He was referring, of course, to the Soviets' success in orbiting a small artificial satellite on October 4, 1957, followed on November 3 by Sputnik II, a launch with a dog named Laika aboard. I think Kennedy was right. In the 1940s, the big events were World War II and the atomic bomb. In the 1950s we had the Korean War and Eisenhower taking over from Truman, but neither of these events had the sudden, decisive impact of Sputnik.

I first heard about Sputnik on my car radio while driving home from RCA. I pulled the car into the garage with a sinking feeling. It's hard to describe the feeling I had on that day. I think it was largely disappointment that another nation had succeeded first. I had given quite a bit of thought to space and satellites for a number of years. Since 1948, I had served in a minor role with the NACA (National Advisory Committee for Aeronautics), the forerunner to NASA. The NACA's subcommittee on automatic stability and control, of which I was a member, openly questioned what the NACA was doing to prepare America for possible activity in space. We had our wrists slapped. We were told that the NACA was for aeronautics, period. Forget space.

Space remained very much on my mind. In January 1953, I gave a talk on the subject, half in jest, to MIT alumni of southern California. I discussed work that I had done with Doc Draper and that I then was doing on Project Meteor. To conclude, I threw in a few thoughts on the possibility of space travel, just to end on a provocative note. People asked when I thought space travel might happen.

I answered, "First a relatively small instrumented payload will go around the Earth."

"Do you think man will ever fly around the Earth?"

My answer was: "Sure, why not?"

"When do you think it might happen?"

"In about ten years." In 1953, my prediction was considered somewhat eccentric. Vostok, the Soviets' first piloted orbiter, made me a prophet.

Before Sputnik, no one seemed to care much about space. After Sputnik, every aeronautical engineer in America had been working on space forever! And the public was abuzz. People were suddenly speculating about Soviet satellites flying overhead. There was fear in some quarters that satellites might be used as platforms for nuclear bombardment, but the more likely threat was thought to be enemy reconnaissance. One Air Force hardliner, with whom I served on a committee, had a drawing showing Russian satellites ringing the earth. He asked at a conference how we felt about this "great web of satellites over our heads." This sort of hysteria was not uncommon, and it made me a little uncomfortable. Nevertheless, I saw a growing concern with space and the Soviets' presence everywhere I looked. While I was still at RCA, we obtained the first contract let directly out of the secretary of defense's office, known as SD-1. Code-named SAINT (for satellite interceptor), its stated purpose was to develop a satellite capable of intercepting, inspecting, and destroying another satellite.

In October 1958, the NACA became NASA, the National Aeronautics and *Space* Administration.[1] I read in the papers about T. Keith Glennan's appointment as NASA's first administrator in October 1958. A Yale graduate, he had served on the Atomic Energy Commission. At the time of his appointment, he was president of Case Institute of Technology, which later became Case Western Reserve University. Within a week of his coming on board, Glennan announced plans to launch a capsule with one astronaut into orbit, a decision leading to Project Mercury, starring astronauts Alan Shepard,

---

[1] Note that whereas the acronym NASA is customarily prounounced as a word ("Nassa"), NACA was always said as a string of initials ("the N.A.C.A.").

Gus Grissom, John Glenn, and other early space heroes. I felt a little jealous of Glennan and the other people who were getting the opportunity to be involved in this exciting new arena. At the time, I had no official ties with NASA, my old NACA committee having been disbanded about a month after Glennan's arrival. Still, I had maintained some of my old NACA ties, and I was invited to serve on a new ad hoc committee on guidance and control, chaired by William H. Pickering, who ran the Jet Propulsion Laboratory (JPL), which later became part of NASA. I also received a call from NASA, asking if I would be interested in moving to Washington and heading up the guidance and control program at headquarters. I turned down the offer as not challenging enough.

On June 27, 1960, I was sitting in my RCA office in Burlington, Massachusetts, when the phone rang. It was Keith Glennan. He asked if I was planning to be in Washington in the next few days. I said I really hadn't been planning to be down there at all, but would be happy to make the trip if he wanted to chat with me.

"Well," he asked, "could you have dinner with me tonight at the Hotel Statler in Boston?"

"Of course," I answered.

At dinner, Keith told me how things were developing at NASA. Then he took a letter-size organization chart out of his pocket, put his thumb down on one of the highest boxes in the hierarchy, and said, in words that were a bit stronger than this, "I'd like you to consider being the associate administrator of NASA."

Keith explained that in NASA's earliest days, he and his deputy, Hugh Dryden, had run the program. (Hugh was the highly respected former director of the NACA.) Glennan and Dryden had soon realized, however, that they needed to have a full-time general manager. "That's a term that generally is not used in the government," he said. "It seemed better to call this person the associate administrator." To fill this position, they had hired Richard G. Horner, the Air Force's former assistant secretary for research and development, who made himself available for one year only.

"Dick's year is almost up," Keith said, "and we're looking for somebody to come in and take over soon. We hope you'll consider it." After he told me more about the job and its responsibilities, I explained to him that there were certain family things I wanted to consider. I told Keith I would be in touch in a matter of days. I went home and talked with Gene. We had already been over this business of moving recently. Three months earlier, my good friend Courtland ("Court") Perkins, then the assistant secretary of the Air Force for research and development, had asked me if I would consider running the NATO (North American Treaty Organization) systems laboratory in The Hague. But on May 1, a U.S. U-2 reconnaissance plane piloted by Gary Powers was shot down over central Russia, and Powers was taken captive as an American spy. This heightened international tensions and may have influenced the NATO countries to select a non-American for the laboratory. At least, that's what Court told me in the aftermath.

It was Court who had recommended me to Keith Glennan, who then checked me out with Hugh Dryden. Gene could see right away that I wanted to take the job, and she assured me that the move would work out well for everyone in our family. Less than a week after my first meeting with Keith, I accepted his offer.

First Weeks at NASA

Once you take on a fairly key government job, there's tremendous pressure to start immediately. Because your name has been mentioned, the organization expects you. I insisted that I could not begin work until the first of September. Keith agreed but said, "There's a three-day industry meeting in Washington at midsummer, and it would be wonderful to have you there. You'll learn a lot, and you can say a few words, so people get to know who you are."

Gene suggested that she come with me and find a house while I was at the conference. This worked out wonderfully. She went house-hunting the first full day we were in Washington. That night she was down to three choices. We finally settled on a new townhouse at 1503 Dumbarton Rock Court, just off P Street in Georgetown. I took a look in the garage and found a brand-new Rolls-Royce inside. The man who lived across the street had two of them and needed the garage space. Otherwise, it was a reasonably modest vertical house costing about $85,000. Gene's mother called it "the Rocket." We had a tiny backyard, which eventually became a garden. We found our neighbors and friends most cordial, as people in Washington are used to families coming and going.

Our September arrival in Washington was far from perfect, however. I went ahead on the last day in August and spent the night in a hotel. The next morning I was sworn in promptly at 8:30. Keith Glennan had suggested that I spend my first month at NASA getting to know the outfit. From NASA headquarters on Lafayette Square facing the White House,[2] I visited the various field offices or "centers," where NASA's research and testing work was done. My second day on the job I visited the Langley Research Center in Hampton, Virginia.

Gene drove to Washington with Kathy and May. Our three boys were still in New England. Toby was at Lenox, while Joe's academic future had been uncertain until the very last minute. He had applied to St. Albans School in Washington, but his acceptance didn't arrive until the day the movers were filling their van in Beverly Farms. If Joe hadn't been accepted, he would have lived with the Lorings in Prides Crossing and continued his studies at Shore Country Day School. He and Daniel were scheduled to fly into Washington the day after Gene's arrival, accompanied by our long-time nurse, Hazel Whitney.

It was terribly hot, and the drive south was quite unpleasant for Gene and the girls, with the three of them crowded into the front seat of our little Fiat station wagon. In one of the more famous Seamans family pronouncements, Kathy, who was at Dobbs and feeling pretty sophisticated, said to May, who was seven and chubby, "Get your sweaty body away from me!" When they finally got to Washington, they discovered that the movers had lost their way, and there was no furniture in the house on Dumbarton Rock Court. So Gene and the girls spent the night at the Marriott. I arrived back from Langley in time to spend the night with them. The next morning I accompanied them to the house and found nothing in it. Daniel and Joe arrived from the airport with Miss Whitney. She was very concerned to find dusty paper on the floor, two dirty spoons and a tumbler in the cupboard, and dust everywhere. How could she feed Daniel, age eighteen months, his cereal? Fortunately, we had some friends in Washington,

---

[2] NASA's offices moved about a year after my arrival to FOB 6 (Federal Office Building No. 6) near Capitol Hill.

Gene's second cousin, Connie Wood, and his wife, Nancy, and they were most supportive. Through them we were able to get what we needed, including mattresses to put on our gritty floor. That first night in our new townhouse, Gene and I slept on mattresses with nothing to cover us but our raincoats.

The next morning I said, "See you all!" Then I walked out the door and jumped into a waiting chauffeur-driven limousine. I was off to visit more of NASA. As the driver and I pulled away from the curb, the movers were pulling in. Gene was left alone to sort out the household.

Tour of the NASA Centers

NASA's centers were a mixed bag. Langley, the first stop on my September tour, was the first aeronautical research laboratory run by our federal government, established in 1917. It was NASA's "mother lode." Several other NASA centers grew out of Langley—among them the Lewis and Ames Research Centers and the Johnson Space Center (originally known as the Manned Spacecraft Center) in Houston. Langley had a number of wind tunnels, some of them large. One was big enough to contain a fighter plane. Some were vertical-wind tunnels, used to test helicopters. With such extensive facilities, Langley attracted exceptionally fine aerodynamicists. The caliber of the work there was always high.

Floyd C. "Tommy" Thompson was the director of Langley. Meet him for five minutes and you would have said he was a waterman or a farmer. When you knew better, you realized he was remarkably effective. One of the top research people at Langley was Robert C. Gilruth, who headed up what was then known as the Space Task Group. This later split off from Langley, moved to Houston, and became the Johnson Space Center. Working for Bob Gilruth was a talented engineer named Max Faget. Together with their team of engineers, Bob and Max came up with the familiar gumdrop shape for the Mercury capsule—a foreshortened cone with a heat shield mounted on its rounded bottom. By the time of my first visit to Langley, the Mercury capsule had been fully designed, and I inspected a dummy version. John Glenn, one of the seven Mercury astronauts, took me in hand, opened up the cockpit for me, put me inside, and closed the hatch, while explaining its workings.

The idea of sending a man to the Moon was still futuristic to most people in 1960, but a few Langley engineers were giving it prolonged thought. During my September visit, a Langley engineer named John C. Houbolt went over his ideas for lunar-orbit rendezvous (LOR) with four or five of us in a small conference room. LOR was one of three considered "modes" for landing an astronaut on the Moon. Direct ascent—in which a rocket taking off from Earth flew directly to the Moon, landed there, took off, and landed again on Earth—was the simplest to imagine, though ultimately it might have proved hardest to pull off. Earth-orbit rendezvous (EOR) was the most popular mode in those days of speculative thought. With EOR, a robot-guided rocket would be launched into orbit, to be followed later by the astronauts in another capsule. Once rendezvous had taken place in Earth orbit, the mission would proceed as in direct ascent.

In the LOR scenario the main capsule achieved lunar orbit before the lunar lander disengaged itself. It landed and then rejoined the main capsule for a return to Earth. One of the critical concerns with LOR was docking. If docking in lunar orbit was not successful, not only would the mission fail but crew members making the lunar landing would be cut off from transportation back to earth. It was thought that a similar problem in Earth orbit would be surmounted easily, since the capsule had the ability to reenter the atmosphere and to land. For me, the LOR docking had a remarkable similarity to the approach RCA had been developing for satellite interception (the SAINT program). But at the time of John Houbolt's informal briefing, few were giving it any real consideration. This mode question would be one of the last major decisions we made on the road to achieving the lunar landing.

Part of the Army's Redstone Arsenal in Huntsville, Alabama, was transferred to NASA on July 1, 1960. This center was as fascinating as any I visited that first month. (Within a month of my visit, President Dwight D. Eisenhower and Keith Glennan would travel to Huntsville and rededicate this part of the Army Ballistic Missile Agency the "George C. Marshall Space Flight Center," after which it was known within NASA as "Marshall or MSFC.") The senior engineers at Redstone, most of them Germans, were working on NASA rocketry under the legendary father of the German V-2 missile, Wernher von Braun. When I got to Huntsville, von Braun was nowhere to be seen, so I began discussing the space program with some of his lieutenants, or "cardinals" as they were known. Then Wernher came in, immediately dominating the scene. He spent time that day and the next showing me around.

Wernher had an amazing presence and made a most favorable impression on me. One seemingly insignificant incident sticks in my memory. When we got out to one of the gantry elevators to go up and take a look at a rocket, there were some workmen waiting to get on. When they saw Wernher, they all backed off. Wernher put his arm around one of the men—a large hard-hatted construction worker—and said, "You're the guys doing the work. Come on. You get in the elevator first." And together we went up in the gantry. I had expected to see a completely autocratic system in the German mold at Redstone. Though it was clear that Wernher thought of himself as the boss if not the Pope, he listened to what his cardinals were saying. Then, and only then, he would say, "Okay, this is what we're going to do," and they did it. He was much more humane than I had imagined.

At an October 1960 NASA management meeting in Williamsburg, Virginia, I got another insight into von Braun's team. I began a little talk by telling everyone that at RCA we had made almost a fetish of calling each other by first names and nicknames. One of the senior vice presidents of RCA was nicknamed Pinky. Though I had trouble calling him that, I finally got used to it. Then I added that I was impressed, while touring NASA facilities, that things there were much more formal, with people addressing one another, especially superiors, as Doctor This and Mister That.

At the cocktail hour afterwards people made a point of calling me Bob. Eberhard Rees, one of von Braun's cardinals, did so with his thick German accent. "Bob, you might be interested to know," he said, "that just the other day, Wernher said for the first time that I could call him Wernher."

"Do you mean," I asked, "that you've been working together for, whatever it is, twenty years, and you've been calling him Dr. von Braun all this time?"

"Oh, no!," Rees answered. "I always called him Herr Doktor von Braun."

I also visited the Goddard Space Flight Center in Beltsville, Maryland, on the road from Washington to Baltimore. Goddard was an offshoot of the Navy that had been given the job of coming up with the Vanguard satellite during the International Geophysical Year, 1957–58. I also took a September trip to the west coast, where I visited the Jet Propulsion Laboratory (JPL). Located very near the Rose Bowl in Pasadena, JPL was a key operation for NASA at the time, with responsibility for Ranger and Mariner, the robotic lunar and martian probes. But it was also a NASA oddball, because it was not civil service but privately operated by the California Institute of Technology (Caltech). All the other centers were run by people on the government payroll. This caused considerable difficulties for Abe Silverstein, director of Space Flight Programs at our Washington office. Being private, JPL didn't like the firm hand of Abe Silverstein on the back of its neck, but there was reason for Abe to keep his hand there. We had about $100 million worth of effort being contracted out of JPL at that time. What's more, Caltech was bleeding off funds as "their fee" to pay for the risk they were taking. When we tried to negotiate these matters with Caltech's president, we were told, in effect, "If you don't like it, lump it. If you want us to run the program, we're going to run it our way." At an impasse, we finally decided that not all lunar and planetary probe exploration would go to JPL, as previously planned. We gave the Lunar Orbiter program, which JPL was expecting to manage, to Langley instead. This was a controversial move. Langley had never run a big project before, and there was a real question as to whether we ought to take our research people and put them on one. But we did it; Langley did a miraculous job; and we finally got JPL's attention.

My September tour of the centers set a good pattern for my years at NASA. I tried to move outside the circle of ten or fifteen people assembled around my office as often as possible. If an issue came up involving Goddard or one of the other centers, I would try, to the extent that time permitted, to get out to Beltsville or wherever the action was taking place. In this way I could understand the issues better while demonstrating that headquarters cared.

Moving the Decimal Point

I was highly impressed with the number of extremely competent people I met in my first tour of NASA. What's more, there was an extraordinary team spirit in the whole endeavor. I won't say there weren't jealousies or animosities, but these struck me as minimal compared with those I had seen in academic and industrial settings. At NASA, there seemed to be a lot less competition and a lot more getting on with national objectives.

Like every other citizen, I had wondered about the inaction and waste of bureaucracies and about the heavy load they put on taxpayers' shoulders. So it came as a great surprise to find that these bureaucrats were, in the main, able, dedicated people. And when it came time to visit Capitol Hill and the White House, I was equally impressed with the level of competence of most everyone I met.

Everything at NASA was much bigger than anything I had ever encountered before. In my fifteen years at MIT, the largest program I had been involved with had a total budget of about $20 million over eight years. RCA operated on a billion-dollar budget, but my program was a tiny piece of that, with an annual budget of no more than $15 million. Now suddenly I was the de facto general manager of a billion-dollar-a-year program, with resources dotted all over the country and countless contractor personnel intimately linked to every phase of it.

In a way, though, it was only a matter of moving the decimal point over a couple of places. Or this was how it seemed to me. World War II had forced quite a few people in my generation to move into big jobs without much prior experience. In the armed services, twenty-five-year-olds had been promoted to full colonel in the field. If it hadn't been for the war, I wouldn't have been an instructor at MIT at the age of twenty-three, before getting my master's or doctor's degree. Ordinarily, you're not considered for a faculty appointment at MIT until you've got your doctor's degree.

Still, at RCA I had never had my boss come into my office and say, "Come with me. We need to go over to the White House to talk with the President and his budget director about a couple of things." And there was no denying the excitement and electricity of working in what would become perhaps the glamor program in American government during the 1960s. I was working shoulder to shoulder with people who were going to orbit the Earth and fly to the Moon.

The Question of Organization

NASA was not all glamor during my first nine months there. The agency had so far failed, in the three years since Sputnik, to catch up with the Russians. We weren't even close. The American public had watched Vanguard and Centaur rockets blowing up on live TV. Only about half of our launches had even left the pad. For good reason, Congress and the public did not have the necessary confidence in the program, and NASA was getting lacerated by the press. We had to be pretty hard-boiled inside of NASA. That's essentially what I was brought in to do as general manager.

When I discussed this with Keith, he said, "We want you to get in there and take charge!" He clenched his fist and made a driving motion, and that was about it. I took this to mean that there was a real need to get on top of the management of the activities at each of the centers. But it wasn't yet clear how to do so.

After about a month on the job, he asked me, "When are you really going to get ahold of this organization?"

I said, "I'm working on it, Keith."

He said, "You'd better move."

This is how NASA was organized when I arrived in 1960. In the Office of the Administrator, there were two people, Keith and his deputy, Hugh Dryden, the only two presidential appointees at NASA. As associate administrator, directly in line beneath them, I was NASA's highest-ranking nonpolitical appointee. Eventually, I had a series of deputies—Tom Dixon, Earl Hilburn, and others. I was slow to name them at first. In response to Keith's promptings, I said, "I don't want to think about a deputy until I know what my job is. Then I can figure out who will be my deputy." It was only after I had been in government for some time that I realized that deputies were not only common, they were in most cases an absolute necessity. The heaving and hauling required of each key individual is so great that it's pretty hard to get the job done without a team of two (sometimes three) to put their shoulders to the same wheel.

Below my office in the line organization were the so-called program offices located at our Washington headquarters. Space Flight Programs, under Abe Silverstein, was responsible for satellites of all types, from meteorological to manned. The Large Launch Vehicles office, under Don Ostrander, had responsibility for all of the booster stages and rocket engines used to launch satellites. Research and Advanced Technology (what amounted to the old NACA) was initially run by Ira ("Ez") Abbott, an old-line bureaucrat. This office continued state-of-the-art research in aerodynamics, propulsion, materials, and so on. The NACA research functions became less and less critical in the scheme of NASA as the nation's space program grew in stature, but we continued to believe that it was important to the country that research continue in aeronautics and long-term space technology.

Beneath the program offices in the hierarchy were the centers, out of which all contracting was done. In other words, the centers were where the work was accomplished. In the organization chart, each center fell under the program office with which it was most closely associated. So Abe Silverstein had Goddard and JPL. Don Ostrander's principal operation was Marshall. Ira Abbott had the old NACA laboratories at Langley, Ames, and Lewis, as well as Edwards Flight Test Center in the Mojave Desert.

Having centers and program offices might seem a redundancy. Why not have the people in the field do the work and forget about headquarters staff? Because an agency has to plan for the future, and it has to defend its programs before Congress, the media, and the public. If people kept coming in from the field to look after these things, work would never get done. There was also a crying need for systems management, a responsibility of the program offices. We had to be sure that the capsule (designed at Langley and in Houston) and the rocket (designed at Marshall) fitted together!

There was always a very delicate balance between the program office and center. After a couple of years I had serious complaints from Harry J. Goett at Goddard, who said, "If I'm running this laboratory, I want to have the power to refuse having people from headquarters come in if I don't want them on my territory. I'm going to be the one to decide what goes on in my center."

That may have sounded like common sense. Only there were fifty projects going on at Goddard, and there was no way that one person could stay on top of them all. The director would be the bottleneck. The people at the program level were complaining that they were not allowed in the Goddard door, and Harry was complaining that he was supposed to be responsible for something while decisions were being made he didn't know anything about!

We finally came to the sad day when I had to ask Harry Goett to come and see me. I had known him since 1948. I admired and respected him, but I had to say, "Harry, you're not able to handle this job." *C'est la guerre.* Harry then became special assistant to the NASA administrator and a year later took a job in the aerospace industry.

Despite its inherent logic, the NASA organizational structure was fraught with contradictions. One example: Abe Silverstein was responsible for Space Flight Programs, but the Space Task Group, set up under Bob Gilruth at Langley to develop manned capsules, officially reported not to Abe, but to Ez Abbott, because Ez had the old NACA centers under him. Another concern was George Low's place in the organization chart. As program chief of Manned Space Flight, he had virtually the most important job in all of NASA, yet he was several levels down on the chart. He didn't even report to Abe directly.

I addressed some of these concerns at lunch one day with Keith Glennan and Hugh Dryden. I suggested that we make George Low director of Manned Space Flight, reporting directly to Abe Silverstein. I also suggested that, in effect, we promote Bob Gilruth and have him report to Abe rather than through Tommy Thompson and Ira Abbott. These steps were taken immediately and addressed a further concern of mine, that manned flight be given top priority. I believed that the future viability and funding of NASA would largely depend on whether or not we were successful in putting an American in space and bringing him back safely.

Still, as for taking charge (in Keith Glennan's fist-pumping sense) of this large, unwieldy network of program offices and centers, I had a hard time. The way the organization was set up, it was difficult for me to exercise my responsibility. There was no system of checks and balances. There was no place for me to grab hold of the total NASA program. It was not until the following summer when, under a new administrator, I took fiscal control of the organization, that I was able effectively to "take charge."

What Will Kennedy Do?

On the first Tuesday in November, John F. Kennedy defeated Richard M. Nixon in the race to succeed President Eisenhower. At NASA, the same questions circulated as at every other federal agency: What is Kennedy's attitude to our work? Will he support or accelerate our programs? Will he cut them in part or altogether? December 19 saw the first successful launch of an unmanned Mercury capsule atop a Redstone rocket. Was this the beginning of something big, or the beginning of the end?

At a postelection meeting with President Eisenhower and his executive officer, Brig. Gen. Andrew Goodpaster, Eisenhower speculated about "what Joe Kennedy is going to try to get his son to do when he becomes President." Eisenhower was unhappy at that prospect, but he was also honorable about it. He said our responsibility was to leave everything in the best posture possible, so that the next administration could take it over effectively.

With this in mind, I requested that contractor studies on the feasibility of human spaceflight, which I had asked for in September, be intensified and extended to include a lunar landing. I also called for a January 1961 meeting at which the Space Task Group under Bob Gilruth and the Marshall Space Flight Center under Wernher von Braun would come in with detailed discussions of a lunar landing—what it would cost, when it might be accomplished, what launch vehicles would be required, and so on. We knew that the new administration would have its own ideas about space. It was our responsibility to have substantive material ready for them, so that the decisions could be made based on facts, not conjecture.

Jerome C. Wiesner, Kennedy's science advisor, headed a transition committee charged with studying what the incoming administration should do about space and the so-called missile gap. I had known Jerry at MIT, going back ten years or so, but surprisingly the NASA organization had no contact with him or his committee during this transition period. When the committee report was issued, we thought it quite unfair to NASA and quite personal with regard to certain key people at NASA, especially Hugh Dryden. It asked for new, young, imaginative people to run NASA, the implication being that Hugh, in his sixties, was old and doddering.

I had lunch with Johnny Johnson, NASA's general counsel, and raised the question of NASA's future. "You know," I said, "we're hearing all these rumors about what the Kennedy administration will or won't do. What do you think is the chance that the next administration will, in effect, dissolve NASA, or reorganize it as a NACA-type operation, shifting more operational parts back into the military?"

"They might want to do it," he said, "but they'd never get it through the Congress." NASA had a lot of support on Capitol Hill. It was seen as a check-and-balance to the Defense Department. Still, even if NASA survived, my own position was tenuous at best. Clearly, I had to be acceptable to the new administrator under Kennedy, and I was quite prepared to leave if requested to do so or if I felt uncomfortable with the new administrator. I think anybody who has a reasonably senior position in the government has to be willing to get out quickly if he or she is not satisfied with the way things are going. Government work is similar to a relay race. The best you can hope for is to move the baton around the track, before passing the responsibility to someone else.

Keith Glennan was a political appointee of the Eisenhower administration. As such, he expected to be replaced by the incoming Democratic administration. He hoped that Kennedy's staff would select a replacement before the inauguration to work with him on a smooth transition. No appointment seemed to be forthcoming, however. After getting clearance from the Eisenhower administration, Keith called the Vice President–elect, Lyndon B. Johnson, to see where things stood on a new administrator. "Here we are," Keith told Johnson, "the sixth largest federal agency from the standpoint of budget, and we're still waiting to have an administrator selected. I just want you to know that I'm ready to do anything I can to help during this transition."

Johnson's reported response was: "That is very kind of you, Dr. Glennan. If there's anything that you can do to help, we'll certainly get in touch with you." With that Johnson hung up.

This was a very difficult period for Keith. At the Wright Brothers Dinner, which occurred in December, he was the principal speaker. Keith tried to make what was probably his last major talk a serious one. The audience paid little attention, many of them talking amongst themselves. NASA was not riding high in the community in those days.

Shortly before the inauguration, Gene and I had a group at the

Chevy Chase Club for lunch. Gene's cousin, Ezra Merrill, president of the Hood Milk Company in Boston, was there. I got to talking with him about this strange anomaly—a sizable agency having no contact with the new administration. He said to me, "You know, if you really want to find out what's going on, you ought to call Charlie Bartlett."

"Who's Charlie Bartlett?" I asked.

"A syndicated columnist and the godfather of the Kennedys' first child." Within NASA, we were desperate for intelligence in those days, so just after the inauguration, I called Charlie Bartlett on the phone and asked if I could come over to see him.

"What about?," he asked.

"Well, frankly," I said, "we haven't had any contact with the administration, and I'm told that you might be able to shed a little light on what's going on."

I met with Charlie and discussed my concerns. He said, "The President is wondering what he might do about the space program. Of course, he's planning to turn it all over to Lyndon. But just last night when I was walking around Lafayette Square with the President, we walked around your building, and the President noted it and said that that's where the headquarters of NASA was located, and that there had been considerable concern amongst his special task force about the competence of the people."

This made me wonder, but at least it showed that the President knew that NASA existed. I said, "Well, I'm sure the President has so many things on his mind. Who am I to say what's most important, but it would certainly be very helpful if in the near term we could have some conversation with the administration."

"I'll remember that next time I see the President." When I got up to go, he said, "Now, you realize that Lyndon Johnson is a very difficult person. He's a real problem for the President. If you have any problem with him, you just let me know and I'll see what I can do to straighten it out for you." I thought, "Seamans, you're getting yourself in a little bit deep on this one!"

On the day before the inauguration, Keith Glennan approved a big list of programs, including a 200,000-pound-thrust, hydrogen-oxygen engine and the so-called F-1 kerosene-oxygen engine with 1,500,000 pounds of thrust, which was eventually used in the first stage of the

Saturn rocket. Then he sat in the office waiting for word on his successor. When no word came through by a quarter to five, a few of us found a bottle of sherry and had a couple of drinks, to toast Keith for the job he had done. Then for the last time he walked out the door—and found himself temporarily trapped in Washington by a blizzard!

Life at the Rocket

Our whole family took enthusiastically to life in the Capital, its opportunities and conveniences, although we all missed family in New England. Our neighborhood proved to be a very pleasant place to live. In time, I went on the vestry of the church around the corner. Right up the hill behind us was Montrose Park, where our two youngest, May and Daniel, enjoyed playing with neighborhood children.

Joe began two years at St. Albans, and May went into the second grade at Beauvoir. These schools, on the close of the National Cathedral, are part of a large Episcopal complex. They offered topnotch schooling and sports, always against the background of religious faith. Joe learned the city with his new friends who lived nearby. May created her own niche and eventually attended the National Cathedral School for Girls, in Washington.

Our youngest son, Daniel, took endless walks. He loved the smelly buses and trucks, and the sirens. He and his nurse would stand outside the houses of notables, especially at the time of the Kennedy inaugural, and wave at the arrivals and departures. The day of the inauguration, January 21, 1961, was an exciting one for all of us. Having grown up in Washington, Gene knew the importance and fun of participating in inaugural events; so we invited my parents to stay with us, go to the parade, and soak up the charged and optimistic atmosphere of the Capital. The blizzard that had hit Washington the day before kept my parents' plane from landing, so it turned back to New York City. Luckily, their fellow passengers included Cardinal Cushing and a covey of nuns. The cardinal was to give the invocation at the President's swearing-in. Realizing the importance of His Eminence, the airline bused passengers on the flight to Penn Station in New York City, where they were put on a special car. There was much jollity en route and, to our amazement, my parents arrived at 4:00 a.m. on inauguration day, very proud of themselves! We had pictured them hopelessly delayed in sixteen inches of snow.

Invited to an Inaugural Ball, Kathy arrived from the Masters School in Dobbs Ferry, New York, where she was boarding. We all had a fine view of the seven-hour parade, then got Kathy off to the party. She was quite a sight as she marched down our snowy walk in her pumps and flowing evening dress held high. Her young man didn't appear very well organized, but he did wear a stunning top hat!

The Coming of Jim Webb

It turned out that the Kennedy administration had been having great difficulty finding somebody to take the job of NASA administrator. They went through quite a list of people. When it became known that people were turning down the top job at NASA, it was not good for agency morale. Why would anybody not want to run NASA? It must mean that the agency doesn't have a high priority in the administration's planning.

Finally, a week after Kennedy took office, Hugh Dryden called me into his office and said, "Well, I've met the new administrator, and we may have a slight change of pace around here." By that point, I was very anxious to meet the new man, whose name, James E. Webb, didn't mean a thing to me. Appointed by Kennedy on January 30, and confirmed by the Senate on February 9, Webb was an old Washington hand. He had been director of the Bureau of the Budget and an under secretary of state in the Truman administration.

Our first meeting in his office was unusual. He asked me a lot of questions but none about space. I have a habit of twiddling the coins in my coat pocket. He noticed this and two or three times said, "Do you want a cigarette? I have some over in my desk." "No," I said, "I don't smoke."

We talked about different management philosophies and styles. He asked if I knew how Sears Roebuck was organized. At Columbia University's advanced management program, which I had attended while at RCA, I had read a case study of Sears Roebuck. The company had a highly decentralized organization with responsibility spread out horizontally. Jim and I discussed this kind of management control as opposed to a straight, hierarchical kind of organization like that at Montgomery Ward.

I felt it was a good conversation, that at least I had been able to keep in phase with the kind of thing that seemed to interest him. Finally, he said, "Well, it's time to go out to lunch. Will you join me?"

I said, "Sure." We put our coats on and started walking out the door of the building.

"Where would you like to go?" he asked.

I said, "I think that should be up to you, Mr. Webb. You know Washington better than I do."

"Well," he said, "what do you say we go to the National Democratic Club?"

"That's fine with me if they'll take Republicans." I wasn't sure he knew I was a Republican, and I figured we had to address the issue sooner or later. For quite a while after that, whenever he introduced me it was as "the Republican member of the NASA organization"— to show how bipartisan we were!

At lunch, he started giving me some of his philosophy on how government works. He told me the story of somebody who had been a hero to him. When asked at the time of his retirement how he had been able to stay in government as long as he had, the hero had answered, "I know when to bend a little." But, Webb added, there have to be thresholds. You have to begin by choosing the point beyond which you won't bend any more.

Webb told me that he really hadn't wanted to return to government, but that he had been persuaded to do so by Senator Robert Kerr (the chairman of NASA's Senate authorization committee, with whom Jim had worked), by the Vice President, and finally by the President. He said he had made it absolutely clear from the beginning that he would accept only if Hugh Dryden stayed on as deputy. By the end of the luncheon, he let me know that he wanted me to stay as well. Apparently, Hugh had urged him to keep me on.

Reorganization

There was no immediate change when Jim Webb took over. He seemed to follow an adaptive course, getting to know the people in NASA and their capabilities, as well as the organization and its needs. He often spoke admiringly of the Wright Brothers' success, noting that they had recognized the need to turn and maneuver their aircraft. Similar adaptation was required in organizations, he said, in order to steer around or over obstacles while aiming at changeable goals. He liked to point out that NASA had both a brake and a throttle. During this period Jim Webb was also getting to understand the political ramifications of what we were doing, and the political support that he could expect to obtain. He took the temper of the executive office as well as the Congress.

As winter passed into spring and Jim developed some confidence that I could provide the support he wanted, I began explaining to him my concern that we had not centralized control sufficiently in the hands of the general manager. He encouraged me to gather in the reins more than I had with Keith. It seemed to me that to do so meant, first and foremost, to take charge of the funding—to approve on a project-by-project basis the funds that would be available to a given office. Under Keith Glennan, I had not had fiscal control. Funding was controlled at the program level, by Silverstein, Ostrander, et al. Silverstein had a first-rate assistant reporting to him named DeMarquis ("D.") Wyatt, who kept track of funding for him. Abe must have realized things were changing the day I said, "Henceforth D. Wyatt works for me." D. did so from that day until the day I left NASA in January 1968. Every dollar NASA spent for the next six years was itemized by D. Wyatt on forms he submitted to me for approval.

As general manager, I did not want random reporting—people informing headquarters when and if they felt like it. A general manager can't sit all by himself and, by reading the reports coming through, figure out what's going on and what needs fixing. A general manager needs to meet face-to-face with those reporting on a regular (preferably monthly) basis. This was my point of view, and it was controversial. But I saw the need for it.

Consequently, we began to hold management meetings for two days each month, during which the project people came in and talked performance, funding, and schedules. D. Wyatt and his staff kept track of project difficulties and successes, the rapidity with which resources were being used up, and the important schedules. They briefed me before I met with a particular program office. Then, when the presentations had been made, if people hadn't brought up things that appeared to be serious in light of the Wyatt briefing, I would bring them up. I would ask, "Is that Eastman Kodak lens going to work or not? What data do you have? How far along are you?" Of course, this process was a two-way street. The project managers were able to share their needs and problems with me and my staff, so that we in turn could make available facilities, people, dollars—whatever was required to get around a problem.

In the organizational structure that evolved under Jim Webb, Hugh Dryden—the man derided as an old fogey by Jerry Wiesner's committee—continued to play an important role. Hugh's handwriting reflected his approach to everything: it was very, very meticulous. Hugh's lifestyle was very much wrapped up in the Methodist Church, where he was a deacon and preacher. He was not a gregarious person, so temperamentally he and Jim weren't alike. They didn't socialize a lot with one another. Yet, although Hugh was a very mild-mannered person, when he was upset, there wasn't any question about it. If he thought NASA was going off on a tangent or getting too political—whatever his concern—he would go in and give Jim a piece of his mind.

Jim sometimes had wild, mind-stretching thoughts. If he had just sprung them on the organization or on people outside the organization without any kind of constraint, it might have been catastrophic. So it was useful for Jim to have Hugh present, as a thoughtful, highly respected senior person to confide in and to use as a sounding board. For example, when the decision was made to go to the Moon, the Draper Lab got the contract for the navigation system. Soon afterward I got a letter from my old mentor, Doc Draper. He said that he had always found it important for people who designed and developed equipment to have a chance to test it, and that he himself was available to go on the first lunar trip. Jim thought this was terrific. As far as he was concerned, he was going to go over and see the President and say, "There may be some scientists in the country that aren't for the program, but here's one that's not only for it, he wants to go."

It would have been highly inappropriate for Doc Draper to go to the Moon. He was over sixty and couldn't possibly have passed the physical or survived the training. It was much more important for him to direct his laboratory and to build the Apollo guidance system than to go to the Moon. If Doc's offer (and NASA's willingness to consider it) had reached the public, however, it would have caused all kinds of interest. Pretty soon a lot of other people would have wanted to go, and choosing the astronauts would have become a messy business pretty fast. Hugh said, "Wait a minute!" or words to that effect, and Jim never took Doc's letter over to the President. I'm not sure I could have put the same constraint on Jim that Hugh did.

Jim, Hugh, and I formed a triad at the top of NASA. Jim was the charismatic leader with long-range vision and a great knack for understanding how policy and politics interacted in Washington. Hugh, the universally respected scientist who had been the director of the NACA in the 1940s and 1950s, possessed a quiet, invaluable sense of practicality. As the "general manager," I managed NASA's programs while Jim lined up outside support and Hugh provided sound guidance on our goals.

Moving Forward

On March 22 at the White House, I met for the first time with President Kennedy. NASA had previously received a request from Kennedy's director of the budget, David Bell, to recommend substantive changes in our budget. We had met with Dave and had discussed several major items, including what we perceived as the imperative need of going ahead with the Saturn second stage, which had been omitted from the Eisenhower budget. We had also discussed the Apollo program (especially the desirability of going ahead with development studies for it) and the need for a government-financed communications satellite program. Bell had not accepted any of these proposals outright but instead had called for the March 22 meeting with the President at Jim Webb's request.

Kennedy and I had been classmates at Harvard, and I had had a speaking acquaintance with him as an undergraduate. He was very cordial when he came into the Cabinet Room, but he did not appear to recall our association at Harvard, nor did I remind him of it. Present at the meeting were Jim Webb, Hugh Dryden, Glenn Seaborg (head of the Atomic Energy Commission), Vice President Johnson, national security advisor McGeorge Bundy, and presidential science advisor Jerry Wiesner. The President asked many questions and listened carefully as he tapped his front teeth with his pencil. He brought up the question of

Alan Shepard's Mercury flight, which was scheduled for later in the spring. He wanted to know whether we thought the flight would be successful. He seemed to recognize that there was much to be gained if we were successful but also much to be lost if we failed. He questioned whether the flights should be completely open to the press or not.

We discussed launch vehicles, with the President trying to determine what restoration of the budget item for the Saturn second stage (S-2) would provide. At one point, I attempted to summarize some of the discussion, indicating that, if we went ahead with the S-2, it would give us an option for preliminary Apollo flights in 1965, circumlunar flights in 1967, and a lunar landing in 1970 or soon thereafter. The President turned, looked at me, and said, "That was a good summary. I would like to have that in writing tomorrow morning. And please send a copy to the Vice President."

Throughout the meeting, Kennedy appeared to be a master at generating discussion and listening carefully to differing viewpoints. But he did not tip his hand. When we left the meeting, we did not know what his decision on NASA's budget would be. I went home from the meeting and did my best to put in writing what I had said at the meeting. The next morning, I took the letter around to Jim Webb and said, "Here's the memorandum requested by the President." He did a double take. Associate administrators didn't write letters to the President! I said, "I believe he specifically asked for it, but it's in your hands. You decide whether to submit it to him."

That afternoon, as I was headed north for some skiing with my family at Mt. Tremblant, I read his cover note: "The attached memorandum prepared by our Associate Administrator, Dr. Seamans, responds to your request of yesterday...." In the end, we got a lot more in the NASA budget than we would have directly from Dave Bell. The Saturn S-2, for example, received the full increase of $70 million we had requested.

Moving Faster

On the day Soviet cosmonaut Yuri Gagarin orbited the Earth, April 12, 1961, George Low and I happened to be testifying on Capitol Hill. This was not my first appearance before a congressional committee. That

had taken place the previous fall, in front of NASA's House appropriations subcommittee headed by Albert Thomas of Houston, Texas. I
had heard stories about him and his committee, and I had approached
this first appearance with great trepidation. I had heard that Dick
Horner had spent most of his time in front of Thomas defending himself. Keith Glennan had had similar experiences. One day, while Keith
read a prepared statement, Thomas went into a phone booth and
waited until Keith was finished. Then he stepped out and said, "That
was excellent, Mr. Administrator. Very fine presentation."

For some reason, Thomas was kind to me during that first session.
I found him a remarkable man, like almost all the congressmen and
senators with whom I dealt. In my opinion, they wouldn't have had
their jobs if they hadn't had some capabilities. Not that these fellows
were paragons of virtue, but I believed it was wrong to sell them short.
Most of them worked hard at their jobs. Albert Thomas, for one, used
to spend a great deal of time going through budget submissions and
underlining them in red. When he came into the hearing room, he was
ready, willing, and able to cross-examine a NASA witness, in a merci-
less way if necessary.

Congressional hearings took on a new tone after the Gagarin flight.
Suddenly, the Hill wanted to know why NASA wasn't doing more in
space. Jim and Hugh testified on April 13 and were somewhat beaten over
the head for not coming up with more imaginative programs. The next
day George Low was on the hot seat, presenting the Apollo program, and
I was sitting beside him. The congressmen asked questions about
specifics—capsules, life-support equipment, and so on. I drew some comfort from the studies on piloted flight that I had requested in September
and had accelerated in November. It was good to have some facts about
the costs of an expanded spaceflight effort. At least when we mentioned
numbers, we could be pretty confident we were in the ballpark.

After a while the discussion broadened considerably. Congressman
David S. King from Utah quoted from the gospel according to Luke,
chapter 14:

> Or what king, going out to wage war against another king, will
> not sit down first and consider whether he is able with ten thousand
> to oppose the one who comes against him with twenty thousand?

Congressman King wondered aloud what the Soviets were planning to do in space and how much effort they were going to put into it. If we didn't know the answers, might we not find ourselves outmanned like that biblical king? I said that obviously I was not privy to the Soviets' plans. At the same time we could see that they had had a very aggressive program in the past, and we could speculate that they were going to continue in an aggressive way.

Were they planning to go to the Moon in 1967, on the fiftieth anniversary of the Bolshevik Revolution? Again I had to say that I didn't know the answer.

How soon were we planning to go to the Moon? I explained that the budget we were submitting would permit us to make a circumlunar flight in 1967 but that it was not paced for a lunar landing until 1969 or 1970.

Finally, Congressman King got around to asking whether we could get to the Moon by 1967. I said that obviously this was a question that couldn't be answered in a definitive way. When pressed, I finally said, "My estimate at this moment is that the goal may very well be achievable." Our preliminary studies told us that it could be done and that we could do it for between $10 billion and $12 billion. Little did we realize that Apollo would ultimately cost closer to twice that.

When our testimony was done, George and I stepped out of the conference room into something that was unique in my experience, the blinding light of television cameras. I was asked to repeat the statements I had made in the hearing room. Asked whether I thought we could beat the Russians to the Moon, I answered, "This obviously depends on how fast the Russians get there as well as how fast we proceed, and I don't think that we can guarantee that with any amount of money that we can get to the Moon before they do. Obviously," I continued, "it's possible to proceed faster if we proceed on a crash program basis. This is a matter of great national urgency to make a decision as to whether this is important enough to proceed with the kind of funding that would be required."

"Mr. Seamans," a reporter asked, "do you feel that we should put our plans to get on the Moon on a crash basis?"

"It is not for me to make a decision of that sort. This is a decision that must be made by the Congress, by the President, and basically by the people of the United States."

When I finally got back to headquarters, I dashed into Jim Webb's office. Nina Scrivener, his secretary, said, "You look a little distraught."

"Well," I answered, "I got in a little bit over my head at the hearing, and I just want Mr. Webb to know that I did the best I could. But it may be that I'll be asked to leave NASA as a result of it all." On page 87 of his book *Toward the Endless Frontier: History of the Committee on Science and Technology*, Congressman Ken Hechler corroborates my concerns of the moment. He claims that my "job was in real jeopardy as a result of the incident." The administration felt that I had gone considerably further than I should have in favor of the lunar landing program. I was getting into a role that was presidential, not one usually played by an associate administrator.

In my defense, Jim Webb wrote a memorandum (from which the following is an excerpt) to the President's special assistant, Kenneth O'Donnell:

> The members of the Committee, almost without exception were in a mood to try to find someone responsible for losing the race to the Russians and also to let it be known publicly that they were not responsible and that they were demanding urgent action so that we would not be behind.... In this atmosphere, we have had to go forward to try to present the material necessary to justify the authorizing legislation for our program as submitted by the President. This meant we had to go back day after day almost as if the hearings were a Congressional investigation of the program itself. We have presented the Program Directors but with Dr. Seamans... introducing witnesses, and stepping into the breach when the going got rough.... In this position, from a reading of the testimony, I believe Seamans has done an exceptionally fine job.... The Chairman and the Democrats have given him little or no support....

Negotiating with DOD

One week after Gagarin's flight, the world learned of the disastrous Bay of Pigs invasion, a failed incursion into Cuba that had been under way for almost three days before it came to light. The Kennedy administration, which had begun with such hope and promise, had been hit with two devastating setbacks within a week. The following day, April 20, the President sent a memo to the Vice President:

> Do we have a chance of beating the Soviets by putting a laboratory in space, or by a trip around the moon, or by a rocket to land on the moon, or by a rocket to go to the moon and back with a man? Is there any other space program which promises dramatic results in which we could win?... I have asked Jim Webb, Dr. Wiesner, Secretary McNamara and other responsible officials to cooperate with you fully. I would appreciate a report on this at the earliest possible moment.

In other words, Kennedy wanted a goal that could move us ahead of the Russians. Like the coach of a football team on a losing streak, he needed a win. Jim Webb and Defense Secretary Robert McNamara were ordered to come in with a joint recommendation to the Vice President (who spearheaded administration space policy).

On April 28, Johnson sent a memo to Kennedy:

> Largely due to their concerted efforts and their earlier emphasis upon the development of large rocket engines, the Soviets are ahead of the United States in world prestige attained through impressive technological accomplishments in space. The U.S. has greater resources than the USSR, etc. The country should be realistic and recognize that other nations, regardless of their appreciation of our idealistic values, will tend to align themselves with a country which they believe will be the world leader. The U.S. can if it will firm up its objectives and employ its resources have a reasonable chance of attaining world leadership in space. If we don't make a strong effort now, the time will soon be reached when the margin of control over space and other men's minds through space accomplishment will have swung so far on the Russian side that we will not be able to catch up.
>
> Even in those areas in which the Soviets already have the capability to be first and are likely to improve upon such capability, the United States should make aggressive efforts, as the technological gains as well as the international rewards are essential steps in gaining leadership. Manned exploration of the moon, for example, is not

only an achievement with great propaganda value, but is essential as an objective, whether or not we are first in its accomplishment. . . .

In less than five months the administration had gone from doubting the value of any human spaceflight (in the Wiesner Report) to calling exploration of the Moon "essential." Still, President Kennedy held off on the Apollo decision partly, I think, because he wasn't sure NASA could deliver. He had been told by the Wiesner Report and other sources that NASA needed new leadership. And he had no proof that we could successfully get a man aloft without blowing him up. For this reason, Alan Shepard's maiden Mercury flight—a fifteen-minute, suborbital mission scheduled for Friday morning, May 5— was critical. We didn't have very much in the way of monitoring equipment in headquarters, but I had arranged to get some real-time information beamed in from Goddard while the flight was in progress. Edward C. Welsh, executive secretary of the National Aeronautics and Space Council, and I sat in the Dolly Madison House with our fingers crossed until we heard that Alan Shepard had been recovered.

In the midst of the extraordinary public jubilation that followed Shepard's splashdown, the President could see two things. One, NASA could manage a program, *mirabile dictu*. Two, sending an American into space had tremendous impact in the world arena. The following morning, Jim Webb, the Assistant Administrator for Planning Abe Hyatt, and I went over to Defense Secretary Robert McNamara's office at the Pentagon. Time had run out on Webb and McNamara. The administration wanted their joint recommendation without delay.

For starters, Jim and I indicated what NASA felt should be done. Our stated emphasis was on a lunar landing. There was quite a discussion of this objective. Some wondered aloud whether this was too short-range a goal. McNamara asked whether the Soviets could proceed immediately to a lunar landing, leaving us announcing a program and them carrying it out within a year or two.

Jim, Abe, and I explained that we didn't think the Soviets had a lunar capability. For one thing, they still had to build a booster, as did we. We felt that we had a fifty-fifty chance of beating them if we targeted the lunar landing. McNamara asked whether we shouldn't embark on a human planetary exploration program. Call me conservative, but I didn't think that we were in any position whatsoever to take that on as an objective. Even now, near the end of the century, it's doubtful whether we could take it on. Thirty years ago there was no question in my mind that that would have been a foolhardy objective for the country.

DOD then trotted out its agenda, which was centered almost entirely around the development of solid motors. Solids need no maintenance while in storage. Liquid motors are cryogenic, meaning they have to be kept cold. What's more, the liquids tend to boil off, so they need to be constantly replenished and usually topped off before firing. For strategic purposes, DOD wanted the ever-ready solid. They also believed that solids were the wave of the future. (The space shuttle would be the first piloted NASA vehicle partly powered by them.) For a variety of technical reasons, we wanted liquids, not solids, for the Saturn. So we agreed that DOD ought to go ahead with solids.

With John Rubel, a deputy director of defense for research and engineering, I was given the responsibility of putting a report into final form for the Vice President. DOD had already written the preliminary version. When I went over it, I was appalled. The report ran counter to much that NASA, and more particularly Jim Webb, was trying to accomplish. It argued that there were too many companies in the defense business. With competition among many firms, the document argued, there was a lot of wasted time and motion in the procurement process. That was absolutely contrary to Jim Webb's view of the world. He felt that competition was essential, as long as the competition was fair. Jim was in the throes of planning for Alan Shepard's triumphal day in Washington on Monday, but I finally reached him Sunday morning. I said, "Jim, we've got a terrible problem with this report. I think it would be much better to start all over again."

"We cannot do that," he said. "I've agreed with McNamara. We'll use the report. It's up to you to work with Rubel to revise it until you consider it satisfactory."

Sunday afternoon, I was at the Pentagon negotiating word changes with John Rubel. I still was not satisfied with the document, and I called Jim Webb to tell him so. He said that as soon as he had had dinner with the Shepards and had made sure that they were properly taken care of, he would come over to the Pentagon. He arrived about 9:30 p.m.

What happened between 9:30 and roughly 1:00 in the morning was one of the great experiences of my life—watching Jim Webb, who previously hadn't had a chance to read this report, start through it page by page with Rubel there and negotiate changes. He'd say to John, "Now, can you really make that statement at this time?" Or "Don't you think it would be better from the standpoint of the public to have it stated this way... ?" Or "If this ever is published in the *New York Times*—of course, we don't expect it will be—but...." Or "Don't you think the President would prefer to have it oriented a little bit this way?"

Thanks to Jim, a lot of things got deleted, and the report contained the ideas that we had agreed to in the first place. By about 1:00 a.m. the job was done. Three secretaries had stayed with us to help with the drafting, and as we got up to leave, Jim Webb said to Rubel, "How are these girls going to get home?" Rubel shrugged, so Jim went out and asked the women. One of the three didn't have a ride, so Jim said, "We'll wait until you're finished here, then drive you."

At about 2:30 in the morning of May 8, we were somewhere in the outskirts of Washington. As we approached the woman's house, the heavens opened. As she started to get out, Jim said, "Relax. No point in your getting soaking wet. Let's just sit here until the rain lets up." Within a few minutes, the rain stopped, the woman got out, and we finally drove ourselves home. It was a great display of southern gallantry.

I was back at the Pentagon by about 7:30 a.m. to take a look at all the retyped pages, making sure the changes we had wanted had been made. Before nine o'clock the document was ready for Webb's and McNamara's signatures. It was submitted to the Vice President that morning. At noontime, I was at a State Department luncheon hosted by the Vice President. When he got up from his chair to leave, I saw that he had our envelope in his hand.

Within days, the White House announced that President Kennedy would address a joint session of Congress on May 25, in what was widely billed as a second State of the Union address. At NASA we knew that spaceflight was one likely topic of this address, but what would the President say about it? His speechwriters, led by Ted Sorenson, were hard at work, and as soon as they had a draft of the space part of his message, it came over to NASA for our review. It called for a human lunar landing in 1967! The price tag was $20 billion, not our estimate of $10–12 billion. (Jim Webb put an "administrator's discount" on our ability to predict costs precisely.) We were aghast. Jim called Ted Sorenson and convinced him and later, the President, that the stated goal should be *by the end of the decade*. In the final version, President Kennedy changed the deadline to "before this decade is out" and said, "No single space project in this period will be more impressive to mankind, or more important for the long-range exploration of space; and none will be so difficult or expensive to accomplish."

More than once during this period, I looked up at the Moon and wondered if we were all crazy. Intellectually, I believed we could do it. Each step seemed to make sense; yet when I grasped the enormity of the job, I wondered. The more we got into planning for the Moon, the more it became something that I accepted emotionally. I believed we were going to give it a good go, and I felt fortunate that I happened to be there at a time when the country was going to embark on a unique undertaking, one that conformed with my own experience and background.

Organizing for a Trip to the Moon

Once we knew what Kennedy was going to say, we didn't wait for the speech. We got going. The question of organization became paramount again. What kind of an office would be set up to manage Apollo? Where would it be located—in Washington or where the action was? What would be the lines of authority? How were the centers to be tied into this office and with headquarters?

We brought in my old boss from RCA, Art Malcarney, with Rube Mettler from Ramo-Woolridge and one or two others who had expertise in such organizational matters. We met in Jim Webb's conference room, where Jim and I presented our thinking about the alternatives, then got their feedback as well as their thoughts on the kinds of people we would need at the senior levels and where we might find such people. Before gearing up for the Moon, we had 18,000 to 20,000 people at NASA; now we were looking at six or seven times as many personnel. Were we going to increase the government operation by that amount, or were we going to limit the growth within the government and go outside to the private sector for most of the growth? In the end, our overall policy was not to increase NASA's size by anything like that number, but—and this was very important—to have the skills and capability within NASA to make the key decisions.

The NACA had bequeathed to NASA a large number of highly respected and competent professional people at Langley, Ames, and Lewis, and we had inherited additional competence from the Army, Navy, and Air Force. We recognized that, for the Apollo program to succeed, we needed to tap these capabilities and, above all, not to destroy them through any ill-conceived reorganization. While we had to tighten down on the management and the discipline of the mission-oriented centers, we wanted to preserve their scientific freedom. Furthermore, Jim was very strongly opposed to having everybody who was working on Apollo going around with "Apollo" emblazoned on their foreheads. He believed the people in the program ought to think of themselves as NASA people first and Apollo people second. Consequently, he did not want the centers reporting to the Apollo director. Instead, they began reporting to the general manager, namely me.

When syndicated columnist Bill Hines got wind of this, he wrote an article calling me "Moon Czar." I was appalled when I saw it. Hines at least was logical. Seamans, he said, was the Moon Czar because in NASA's new organization the only place key decisions could be made was in Seamans's office. Surely, then, Seamans was the person responsible for taking us to the Moon. There have been few other times in my life when I have thought, "This is so far beyond what I ever thought I'd be doing that it's frightening."

At this time, we were looking for someone outside of NASA to take charge of day-to-day management of the Apollo program. When I came into NASA, all of the space projects had come under Abe Silverstein. Abe certainly wanted to run Apollo, but the chemistry would have been very difficult between him and von Braun in Huntsville, given their very different personal and professional backgrounds. NACA people like Abe were skeptical of von Braun and viewed him as a big spender and self-promoter.

Jim Webb and I started out thinking that we would try to get Levering Smith, who was running the Polaris program. We heard he was dissatisfied with having been a Navy captain for many years. When we floated that one at the Defense Department, he was made an admiral in about ten days!

What about bringing von Braun into headquarters and having him run it? One rather late night Jim Webb and I got talking about this possibility. Why couldn't he run Apollo? After all, he had had considerable success at Peenemünde, the Third Reich's secret rocket lab. The more we talked, the more infatuated we got with the idea. The next morning Hugh Dryden got together with us. Jim turned to me and said, "Bob, why don't you tell Hugh about our recent thinking on running Apollo?" Hugh sat there for what seemed like an eternity (probably five seconds) and said, "Well, if you and Jim want von Braun, that's fine with me. I'll take early retirement." That was the end of our thinking on that possibility.

One summer night at a lawn party at Jim and Patsy Webb's house, I asked Jim and Hugh if either of them knew Brainerd Holmes. Neither did. I explained that Brainerd had responsibility for the biggest project that RCA ever ran, the ballistic missile early-warning system (BMEWS), with tremendous technical and logistical difficulties. They were intrigued, and the following day Jim called Art Malcarney, Brainerd's boss. When Jim asked Art's permission to talk with Brainerd, Art reluctantly agreed.

About three days later Brainerd Holmes met Jim and me at the Metropolitan Club for drinks and dinner. Jim was at his mind-stretching best, pointing out the importance of the Apollo program to the nation and the world, and explaining the role that Brainerd would play in this grand production. Brainerd tried to argue how important he was at RCA. Halfway through the dinner Brainerd turned to me and said, "This boss of yours is really something, isn't he?" Brainerd agreed to think about Jim's offer, and about three days later he accepted.

With Brainerd Holmes coming in, there was no place at the top for Abe Silverstein. One day, he came into my office and said, "If I'm not going to run Apollo, what I would like to do is go back to Lewis as its director."

I told him, "Abe, you've got a handshake on it. If that's the way it comes out, I'll be delighted." That is the way it came out, and Abe was a perfect director for Lewis.

Brainerd (and others) believed that he had to acquire a systems competence. Addressing this need, Mervin Kelly, former director of the Bell Laboratories and special consultant to Jim Webb, told us about Joe

Shea, who had worked at the Bell Labs before moving out to the west coast. Joe was brought into NASA to hire a systems group in Washington, but after two or three months, he hadn't put together a systems team. It's very hard to get a large number of top people to sacrifice a job in private industry that probably won't be held open for them, to take a significant pay cut, and to enter government work, unless they're being asked to fill top jobs. So we decided to set up something outside of the government. I was given the duty of talking to Jim Fisk, director of the Bell Labs, about setting up a systems lab. The resulting systems lab became known as Bellcomm. In a way this created a redundancy. The centers now had two parties looking over their shoulders: the program office to which they reported and the systems integrators at Bellcomm. This didn't bother Jim Webb. While recognizing the need for a line organization in order to get things done, he felt that there should be a multitude of paths through which information flowed to those in charge.

The interesting thing about organizations is that unlike scientific experiments, you never can repeat them to see how they would have come out if you had set them up differently. There is no perfect way to run programs such as those undertaken at NASA. It's not just how you draw the organization diagram, it's the people involved that are important as well. Still, the issue of organization is fundamental to success on large-scale projects.

Long Hours, Hard Work

Of course, while we were gearing up for the Moon there were current programs that needed managing. In the summer of 1961, the Mercury program still hadn't achieved orbital flight, and we had to make sure that it moved along. At the same time, we recognized that very little could be gained from Mercury, other than the experience of putting an astronaut in orbit for a day or so. There was no room in the Mercury capsule for anything that would permit the capsule to maneuver in space or to perform other activities relevant to a Moon mission. Mercury proved to be difficult to turn off, however. The astronauts all wanted it extended because they could see that Gemini, the two-man orbiter, was at least a couple of years away.

This question came up specifically when I went down to Florida to greet Gordon Cooper, the last of six Mercury astronauts to fly, in May 1963. At a press conference, the question was asked, "Will there be another Mercury flight?" I answered that it was extremely unlikely, that it was time NASA moved ahead on the Gemini project. The astronauts and engineers who had been working on Mercury didn't like this answer at all, and I became their public enemy number one for quite a while.

In the robotics area, we had had several failures with the Ranger, a craft designed to make a "hard landing" on (bomb into) the Moon. Then there was the Surveyor, the soft-landing robotic lunar vehicle. It was clearly very important that we learn something about the lunar surface before sending astronauts there. There were many worrywarts, Tom Gold of Cornell University being the most outspoken, who said that the surface of the Moon was covered with dust. He thought the astronauts could well step out onto the lunar surface—and sink from sight! Others said the Moon must be "just like Arizona." Who knew?

Needless to say, there was a lot of work to do. In a nutshell, my day-light hours at NASA were a series of letters, phone calls, and meetings. We had "project status reviews" two days every month—plus time ahead with the people who were putting the agenda together and time after-wards to review the decisions and implications of the meeting. Once a month, on Saturdays, we had all-day "program reviews." In the course of a year, we would attempt to cover every aspect of every one of our programs in considerable depth, for the purpose of bringing the administrator and all of our key people up to date. The following Monday, almost the same presentations would be given for the benefit of key people from DOD, the White House, and Congress.

Then there were formal meetings with Mr. Webb and Dr. Dryden when we were considering procurements. Any time NASA procured bil-lions of dollars worth of rocketry, there had to be a detailed plan for doing so. We had to figure out who the potential bidders might be and which ones were competent. We had to provide a schedule and enough information so that the bidders would know how to bid. How many development models were required? How were tests going to be conducted before designing the real articles? Would the procurement be cost-plus or fixed-price? Or would it be an incentive contract? All these questions had to be answered in the procurement plan.

Then a source-evaluation team had to be selected, one that

would be a cross-section of NASA. (In some government circles this is called "source selection team," but Webb insisted that the selection be made by the administrator. He never wanted to be locked into a decision.) Any given contractor might submit a stack of reports and papers two or three feet high. The source-evaluation team had to go through all of it.

Once the team had prepared their findings, Webb, Dryden, and I would sit at the head of a table and the team would make their presentation. If the project was something coming out of the Marshall Center, Wernher von Braun would be there watching, though he would have no say in the meeting. Our procurement people would be there as well. Webb used such meetings as a way of educating NASA, as well as a way of looking for hidden agendas. If anybody was trying to steer the project toward a particular contractor for whatever reason, we would try to smoke it out.

Afterwards, the three of us would go into Webb's office with our chief procurement person, Ernest W. Brackett, and with Wernher (or his counterpart from the interested center). "Okay," we would say, "we've heard what the source-evaluation people came up with. Now we'd like to hear from you, Wernher. What wasn't considered? Is there anything that was left out that you feel is important?" When he and Ernie Brackett had had their say, they would leave, and the three of us were left with the decision.

As the junior person, I always went first—"Okay, Seamans, how do you look at this?" We would discuss it back, forth, and sideways, as Hugh and Jim advanced their views, too. Finally Webb would say, "Okay, whom do you think we ought to pick, Bob?" I would tell him and why. Then Hugh would have his turn. The next morning Webb's exec would have prepared a one-page decision paper, which said the administrator of NASA had selected, say, North American Aviation for negotiation for the second stage of the Saturn and gave reasons for the selection. All three of us would sign it. The press release would be based on this document, but the document itself was kept on file at NASA in case there was ever a congressional investigation.

We had to put budgets together, then meet with the Bureau of the Budget (BOB)[3] to get them approved. (Federal budgets occasionally were balanced in those days!) The officials at BOB were always trying to hack

---

[3] Now the Office of Management and Budget (OMB).

away at our budget. As a taxpayer, I was glad they did, but as someone who needed money to get a job done, I took a different point of view. As we got closer and closer to putting the final budget together, the process became more and more excruciating. There were some items that had to be taken up with the President. Finally, the budget went up to the Hill, where it had to be presented to both the House and the Senate. That took a lot of time.

On the day the new budget was released, the media attacked. All over Washington correspondents nosed into the different agencies to find out what was important in each budget. We always had a press conference, which I conducted. As many as a hundred correspondents would be present. There might be TV, if the subject was sexy enough. I would run through what was novel in the budget, a process that might take two hours. Then there were questions.

I got a fair amount of scar tissue from my years in government, and a fair amount of that came from the media. The intense media interest in the space program was a shock to me. I liked working with many members of the press. I understood that I could get gored, but I tried my best to have a good relationship with them. Most of them were pretty interesting people and fun to chat with, but I had to be very careful.

Bill Hines, the syndicated columnist who had called me "Moon Czar," was particularly brutal on NASA. He would stand up and fire questions at us in a nasty, incisive way. Why were we so plodding? Why weren't we moving faster? Why weren't we more imaginative? When I came home Thursday nights, Gene would not let me read his syndicated articles until after dinner. Or she served me a martini first, which helped some.

I remember asking John Finney of the New York Times, "Why can't you do a positive, upbeat kind of story on NASA once in a while?"

His answer was: "Okay, I write a good article and if I'm lucky it will be on page 33. If I write something controversial, I have a chance of getting it on page 1. It's as simple as that. I'm paid by what page I get my articles on."

Just before I left NASA, we launched the first Saturn V at the Cape. I went down for the launching. It was a big deal. We must have had a thousand members of the press there, including some Soviet journalists. We got some terrible questions, and afterwards I said to

Kurt Debus, director of the Kennedy Space Center, "That was pretty rough. I'm sorry you have to go through this kind of bearbaiting."

He said, "It is rough, but you've got to realize that during World War II, we didn't have any competition in the German press. The story was told one way only, and I came to believe whatever Goebbels told me. This is tough, but it's a lot better than a controlled press."

NASA was phone calls; NASA was meetings and conferences; NASA was mail. The mail alone was a huge production. An exec in my office would determine who in the organization might need to see a piece of my mail before I did. By the time a letter got to my desk, it might be in a big folder attached to memos from two dozen people in the organization, all of whom had ideas about how I should respond. Then when my draft response had been written by me or by my assistant, there might be others I would want to review the response before it went out. So it might be recirculated with an endorsement attached, and each would initial his approval.

Of course, I got some personal letters, which I was first to see, like those from John Houbolt. John was the engineer who had briefed me and others on lunar-orbit rendezvous (LOR) during my first visit to Langley. In May 1961, he sent me what he called "a hurried non-edited and limited note to pass on a few remarks about rendezvous and large launch vehicles." He said he was for much greater effort on rendezvous and found the launch vehicle situation "deplorable." I answered him courteously. In November 1961, he sent me a second letter, nine pages long, in which he described himself as "a voice in the wilderness" on the LOR question. I found it disconcerting. He may have been right about the mode question, but you can be right and still be courteous. My first reaction was, I'm sick of getting mail from this guy! I thought of picking up the phone and calling Tommy Thompson, Houbolt's superior at Langley, and telling Tommy to turn him off. Then I thought, "But he may be right. We've got to be sure we are considering this alternative. The organization may not be very keen about it, but it makes a lot of sense to me." So instead of doing what my emotions told me to do, I had the common sense to take the letter to Brainerd Holmes. "I've got another one of these zingers from John Houbolt," I told Brainerd. "I'd like to have you read it while I'm here."

He read it and grimaced. Then he said, "Well, we really are looking at

LOR, and it does seem to have a few benefits that the other modes don't."

I said, "I'd really like to get a thoughtful response to the possibility of this mode within the next couple of weeks." A couple of weeks went by, and I hadn't heard from Brainerd. I called him on the phone and said, "Where's that letter?"

"We're seriously considering it," Brainerd said. "It really starts to look like the way to go. Bob Gilruth is sure it is. We can design a lunar lander to be used only for the landing. It never has to return to Earth, so it never has to reenter the atmosphere. So it can be made lighter, without any heat shield."

That, of course, is just how Americans finally got to the Moon and back. On November 7, 1962, Webb announced NASA's choice of LOR and of Grumman Engineering as contractor for the lunar module. As Charles Murray and Catherine Bly Cox put it in their fine book, *Apollo: The Race to the Moon*, "Eighteen months after the nation had decided to go to the Moon, NASA had decided how."

Bringing the Work Home

One Christmas week during the NASA years, everybody in our family was in Beverly Farms except for Joe and myself. The two of us were planning to fly north together on December 23, but when the weather soured we got tickets on a noon train out of Union Station instead. Before leaving Washington I had to go into the office. Joe came with me and sat in my outer office for a couple of hours. There was a lot going on. The phone was ringing, and people were rushing in and out. When we got in the car to be driven to Union Station, Joe said, "I can't believe it, Dad. Is this the way it is every day?"

"Well, it's not quite this bad," I answered. "When people know I'm going to be away for a few days, the office is always more hectic as people try to tie up loose ends and settle matters that need my attention." Still, we did have a great team of people working very, very hard all the time.

Like so many things for my generation, hard work and a belief in it went back to World War II. One year during the war, Doc Draper said he thought it would be a good thing if we took Christmas Day off! There were times when I would go to work in the morning, work through the night, then work all the next day in order to get something out on time. My generation built up a do-or-die work ethic. It amazes me when I look back.

Unfortunately, such a busy professional life leaves little time for family. Each night when I came home from NASA headquarters, Gene would look to see if I was carrying two briefcases of work or just one. My policy was that whenever I could I would bring the work home, instead of staying at the office until the wee hours. The Defense Department operated differently under Robert McNamara. He expected his senior people to be at the office when he arrived and to stay until he left in the evening, period. He got to work at 7:30 and worked until 9:30 or 10:00 at night. To me, that's a terrible way to operate.

I preferred being in the house with the family. I had to bring stuff home every weekend, but at least by being home I could have meals with Gene and the children. "How about this afternoon?," Gene would ask. "I've got an awful lot to do, but why don't we take a walk from 3:00 to 3:30?"

Now that they were growing up a lot, the children fortunately were not especially conscious of these pressures. Kathy was graduating from Dobbs and finding her way to Stanford University. Toby was finishing up at Lenox, while Joe was boarding at Phillips Academy in Andover, studying hard, learning photography, and greatly enjoying crew. His team, of which he became captain, eventually went to the Henley Regatta, where they were beaten by a canvas in the final race. May, meanwhile, was swept up with school, friends, riding, and playing the cello. Daniel went to a school, the National Child Research Center, that experimented with new methods of teaching and equipment. Summers were spent at Sea Meadow, a blessing for us all, and I got away from Washington as much as possible to be with Gene and our children there. In the winters, we had many guests in Washington and plenty of visiting family.

The Politics of NASA

When I arrived at NASA, a Convair was used to fly people to meetings. On the plane there was a steward, who had liquor available for purchase. The first time Jim Webb took the plane, he was offered a drink. Almost the minute he got back to headquarters, he called me.

"Come on into my office. We've got a problem to solve." When I was face-to-face with him, he said, "Take the liquor off that plane."

"Are you against drinking?"

"No," he said, "but we can get into terrible trouble with the Hill if we serve liquor. How does it get paid for?"

"Didn't you pay for your drink?," I asked.

"I wasn't going to have a drink in the plane. Where does the money come from?"

I explained, "It's a revolving fund."

"Well, where did the money come from in the first place?"

To be honest, I didn't know. "I'll tell you tomorrow," I said. A day later I reported, "We got the service going by bootstrapping money from the Langley cafeteria. It is government-subsidized but people pay for their food when they go there."

"Take the liquor off the plane."

I said, "You're going to have a lot of unhappy people."

"They'll be even less happy if Congress climbs on us for something as silly as that."

The next time I saw him I said, "I've got an idea, Jim. What would you think if you and I both put up a hundred bucks and stocked the plane. Then we'll make sure we charge enough for liquor so that the fund is perpetual."

"Okay," he said, "I'll do it."

Not too long after that, the Space Task Group was moved from Langley to Houston. With all the travel back and forth, the old Convair chugging along at about 200 miles per hour was far from ideal. Here we were in the space age—preparing to send astronauts to the Moon—and we were flying around in an old clunker. I could see that we needed a jet airplane. This was a new issue to bring to Jim Webb.

"We're not getting any fancy jets in this organization!," he said. "As soon as you do that every congressman who is involved in our program will want to borrow the jet. No jets!"

So it was back to the drawing board. I finally came in with the suggestion that we get a Gulfstream I, a propeller plane with a jet engine. It was slow enough compared to Air Force jets that no congressman in his right mind was going to ask for it when he had access to Air Force planes. We bought four, with one stationed in Washington, one at Marshall, one at Houston, and one on the west coast. Each carried eleven people. They went about 350 miles per hour—not fast but a lot better than the Convair. Thirty years later, NASA personnel are still flying aboard the Gulfstream I.

As these examples demonstrate, Jim Webb was a consummate political animal, who saw the political implications in everything. He spent a lot of time getting a feel for the viewpoints of senators he felt were important on the Hill, not only those on our space committees but those with close ties to the military. Obviously the administrator of an agency doesn't want to reach the point where he is beholden. Too intimate a relationship with the Congress can tie his hands. On the other hand, anything he does has to be authorized by the Congress.

Through Jim I got to know Senator Robert Kerr, who had taken over the Senate authorization committee for space from Lyndon Johnson. When the decision was made that the Space Task Group be transferred to Houston, in Congressman Albert Thomas's backyard, Kerr had some memorable advice for Jim Webb. "Okay," I recall him saying, "you've made the decision. [Congressman] Albert [Thomas of Houston, Texas] is going to want to have everything he can possibly get. He's going to want to see that center grow and grow. But be sure you don't give it to him all at once. Make sure that every time you add something in Houston, you exact a price for what you need some- where else, for example, JPL." (Thomas was dead set against much- needed funding for the Jet Propulsion Laboratory.) This was a new experience for me, and I found it absolutely fascinating.

In the summer of 1961, Kerr indicated to Jim that when the autho- rization bill came to a vote, he would appreciate it if I could be there with him on the floor of the Senate. Jim had made a point of calling NASA bipartisan and loved pointing out that I was Republican. In this case, it was thought that my presence on the floor would be more beneficial than having Jim there. Senator Kerr had picked the afternoon of baseball's All- Star Game for the vote, so there wouldn't be many people there to con- test the bill. I sat at a desk beside him. (At one point, I lifted the cover of the desk and saw Harry Truman's name carved in it.) There were very few people on the floor when Senator Kerr proposed that the bill be passed. All of a sudden, Senator Proxmire came tearing in with a great stack of papers—amendments limiting funds for JPL. (Thomas wasn't the only lawmaker who disliked the Pasadena lab.) Proxmire began to speak. He started in a measured way, but the longer he talked, the wilder he got.

When Proxmire had finished, Senator Kerr asked me for ideas as he stood up. I told him we needed more housing, the colloquial term for more office and laboratory space. Kerr gave one of the greatest speeches I've ever heard, about the poor scientists and engineers at JPL who haven't proper accommodations for their families, and so on. As he was talking, I tugged at his coat. He leaned down a little bit, and I said "laboratories, more space for laboratories." Without breaking stride, he went on about the need for more laboratory space.

Eventually, it was time for a vote on the amendments. The gongs rang and everybody came rushing in. A few of them went over to Proxmire, but most of them came to Kerr. Some asked, "We're against this amendment, aren't we, Bob?," and things like that. Kerr answered, "Yes, absolutely." Some of them asked a few questions, and once or twice Kerr said, "Bob Seamans can answer that for you." When the vote was taken, Proxmire had no more than a dozen supporters.

Brainerd Holmes Moves on

By the summer of 1962, Jim and I knew we had a problem with Brainerd Holmes. No question, Brainerd was a very exciting person for the media. He had a way of expressing himself that made news. He was determined to move forward as fast and aggressively as possible on the lunar flight, but he had difficulty visualizing the totality of NASA's program, one of Jim Webb's pet objectives. On at least one occasion, Brainerd told me, "I don't understand what the boss is talking about a lot of the time on these general things, and I couldn't care less. That's not my job."

Brainerd made his attitude pretty clear throughout the agency. He expected full support for Apollo from everybody else, but he wasn't about to help on matters other than Apollo. His people liked him. They thought, "It's wonderful to have a gung-ho leader!" And his people began adopting his attitude. At an organizational meeting at Langley, some of Brainerd's people raised disturbing questions about why we weren't giving Brainerd more authority.

There were budget battles with Brainerd throughout the latter half of 1962. At one meeting, a question arose on the allocation of certain funds. We were faced with a reduction in our request, and I wanted to prorate it. This would have meant a slightly larger reduction in Apollo's budget than the congressional reduction called for. Brainerd raised a fuss, getting into a terrific argument with Ez Abbott. That night Ez called me on the phone and said, "I'm quitting." He may have been of a mind to do so before this, but Brainerd's outburst was the last straw.

Jim and Brainerd got into a battle over whether or not to request a $400 million supplement to NASA's 1963 budget. Brainerd leaked to the press that there was a rift between Jim and himself over the supplemental. He let it be known that either he or Jim would probably have to leave the agency over this and that the one to leave would not necessarily be Brainerd! When the President read of this in one of the big national weeklies, he immediately called a meeting with Jim and Brainerd. This meeting was held at the White House on November 21, 1962.

Jim asked me to present his view (and mine) at the meeting. The President turned to Jim and asked how long the presentation would take. Jim answered, "Mr. President, Dr. Seamans will take about thirty minutes."

"I have fifteen," the President retorted. Kennedy did not like formal briefings!

I started in, realizing I had to proceed a little faster than planned or else leave something out. After a couple of minutes the phone rang and the President said, "Excuse me a minute." He turned around in his chair and spoke into the phone, "Yes, Mr. Speaker. . . . No, Mr. Speaker. . . . No, Mr. Speaker. . . ." When the conversation was finished, Kennedy turned back to me and said, "Excuse me, go on." When I had completed my presentation, the President said, "Now I understand that Mr. Holmes wants a supplemental in order to speed up the program and perhaps go to the Moon before 1967. Is that true?"

I said, "Well, I think we might like to have additional funds, but I would very much doubt that we can go to the Moon in 1966. As a matter of fact, we would really need those dollars to hold the 1967 date."

The President asked Jim why we couldn't reprogram funds from other projects, thereby providing the $400 million for Apollo. When Jim demurred, the President said he was not certain that he and Jim were on the same wavelength. "What is our principal objective in space?," he asked. Jim answered that it was to be preeminent. The President then asked for a letter explaining NASA's views, to be delivered promptly. Hugh and I wrote for Jim's signature a lengthy statement of the Webb-Dryden-Seamans position, which concluded: "The manned lunar landing program, although of highest national priority, will not by itself create the preeminent position we seek." The letter highlighted, in addition to the Apollo effort, NASA's important work in four areas: space science, advanced research and technology, university participation, and international activity. The question of a supplemental never came up again, leaving us to believe that the President recognized that there was more to our space program than landing an astronaut on the Moon.

Things went grinding along with relationships getting worse. Among other provocations, Brainerd took the NASA coding structure for handling funds and put it through a conversion matrix in his own office, so it was difficult to track funds when they went to the various centers and out to the contractors. He also called for the firing of Bob Gilruth as director of the Manned Spacecraft Center, a move Jim, Hugh, and I vehemently opposed.

Following the final Mercury flight in mid-May 1963, we had a luncheon in the big state dining room on the eighth floor of the Department of State building. Jim Webb wanted to use the occasion to thank the many groups that had helped us with Mercury—the Air Force, the Navy, the weather forecasters, the contractors, and so on. This was the theme of the luncheon, and I thought Jim presented the accolades most effectively. I went home that afternoon, where I got a call from Brainerd Holmes. He was very upset, to put it mildly, that he had received no recognition.

On June 12, 1963, Brainerd resigned. The press made a big deal of his departure. On July 13, the *New York Times* ran the headline "Lunar Program in Crisis" over a story about Brainerd's resignation and a perceived cooling of public interest in Apollo. I think the press made too much of the whole thing. When I was in Italy a year later, someone asked me how the battle was going between Jim and Brainerd!

Another Wedding

The crisis of Brainerd's last year at NASA coincided with what seemed at the time like a crisis in our family life.

Kathy was always the most high-spirited of our five children. I'm not sure Gene and I handled her joie de vivre very well. When we were living in Beverly Farms prior to the move to Washington, she had a tremendous number of friends. Buzzy Burrage, who was about a year older, would come flying down our driveway on a motorbike; she would hop on sidesaddle behind him with a great big straw hat on; and they would fly back up the driveway, accelerating all the way. Hold your breath! She would try anything.

Kathy's high spirits were matched by high intelligence. In her senior class at the Masters School in Dobbs Ferry, New York, often called Dobbs, four girls applied to Stanford. Only Kathy got in. Now all of a sudden our high-spirited, very smart daughter was 3,000 miles away at Stanford. Great! Gene and I went out to visit in the spring of her freshman year. When we offered to take her out to dinner, she asked if we minded including Lou Padulo. Lou had just received his master's degree and was teaching at San Jose State. He also had charge of one of the fraternity houses at Stanford. I asked, "Didn't he used to work at RCA?" He did. Although we did not want to share our daughter with anyone, we agreed to take Lou along. He was clearly the man of the moment.

The following January we got a call from Kathy on a Sunday night. She didn't call often, but we loved hearing from her when she did. "Hey, guess what?," she said.

"What?"

"I'm engaged!"

"Well, that's wonderful, Kath. Who is it?" It was Lou.

The following Monday morning, I was scheduled to begin three days at Caltech. I mentioned this to Kathy and said, "So Wednesday afternoon on the way home, why don't I swing around and have a chat with Lou?" There was silence at the other end. She finally agreed to the meeting.

Tuesday, while at Caltech, I got a call from Gene. She was very upset. "I've just heard from Kathy," Gene said, "and she and Lou got married in Reno yesterday."

"Well," I said, "I think I can get up there by two tomorrow afternoon."

"I'll be there," Gene said.

I called Jim Webb to explain the situation and to say that I would be leaving Caltech earlier than expected. He said he knew this was something I would want to deal with carefully. "If there's anything I can do," he offered, "let me know. Just feel that the full resources of the U.S. government are at your disposal." He meant well!

When I arrived at the airport, Kathy was there to meet me. We waited for Gene; then after a long chat, the three of us drove to Palo Alto to see Lou. There was a lot of sorting out to do. Gene and I both had to get used to the idea of Kathy's being married and of Lou being our son-in-law.

After two days together, the four of us left things in a somewhat unsettled state. Gene and I came home to tell the rest of the family about it. Two or three weeks later we got a call from Kathy, who said she wanted to come home and visit with us. She felt they had made a mistake, and they now wanted some kind of family service with the two families getting together. She wondered if there weren't some sort of service that would bless the marriage. We said we didn't know, but if so, it would be great by us.

We went to Canon Martin of the Washington Cathedral, who was also Joe's headmaster at St. Albans. He told us we could have such a service—one that was so close to a bona fide wedding service some witnesses might not even know the difference. He said he would love to perform it for us, but wouldn't do so until he had met Kathy and Lou. "If I don't think it's a good marriage, I won't do it," he said. He met with Kathy on a Saturday morning, and they spent about three hours together. People kept coming in to see Canon Martin while he was meeting with Kathy, and he turned them away. He devoted a lot of time and thought to her situation. When they were finished, he said he would be happy to perform the service.

It became a great family event. The Howard Johnson's across from the as-yet-unbuilt, soon-to-be-notorious Watergate complex had itself just been built. We took over the top two floors. A seven-year-old in our party set off a fire alarm. Fire engines came roaring. It was quite a time. We had a reception at our house on Dumbarton Rock Court, attended by Jim Webb and a few special friends. The whole affair was a success. In retrospect, I think the young couple's great concern (and the reason they were reluctant to tell us about their relationship) was that his family and ours seemed on the surface to be worlds apart. As it turned out, we had a lot in common, and both sets of parents valued similar principles.

Not long afterwards, Lou received a NASA fellowship to go to Georgia Tech to get his doctorate in systems electrical engineering. I had nothing to do with his receiving the fellowship, though I certainly was pleased to get the news. I was also pleased that Kathy decided to transfer to Georgia Tech as a sophomore and changed fields from languages to mathematics. Lou had told her there was no money in languages and encouraged her to develop her proven aptitude for math. When she graduated in 1968, she was one of only twenty-three women in her class and one of a handful in sciences; and she had been elected to Tau Beta Phi. Our first grandchild, then about eighteen months old, sat in my lap at her graduation and shouted, "There's mamma!," as she paraded past us in cap and gown.

From what some considered a questionable beginning, Kathy and Lou have made a great success of their life together—and have helped many students along the way. From 1978 to 1981, Lou and I were deans of neighboring engineering schools—he at Boston University and I at MIT. We attended a number of national deans' meetings together with our wives. I came to admire his foresight and courageous style. While living in Brookline during this period, Kathy refinished her house in such a way that six international students could board with them. Each student was responsible for procuring and preparing one dinner each week. The students ate the meal at the kitchen table with Kathy, Lou, and their two sons, and many became lasting family friends.

George Mueller Moves in

By early July 1963, NASA still had not found a replacement for Brainerd Holmes. I was looking forward to my vacation. Caleb Loring, our neighbor Sam Batchelder, and I had entered *Serene*, which we jointly owned, in the Halifax Race. After the race we planned to sail back as far as the Maine coast. The week before I left

Washington, Jim said, "Our bill has got to go up before the Senate this next week. Couldn't you put off your vacation?"

"We can't put off a race in which eighty-five boats are entered," I said, "but if you think I should, I can pull out of it."

"No," he said, "you've planned this, and it ought to be possible to arrange things in NASA so that you can get a vacation now and then."

Racing with Caleb, Sam, and me was Eddie Parker, the navigator. Gene and Eddie's wife, Nat, joined us in Halifax and cruised back to the coast of Maine with us after the race. We came out of the fog into a remote, idyllic place called Roque Island, Maine, where thirty or forty boats were getting together for a day-long rendezvous on a sandy beach. About two minutes after we broke out of the fog, somebody hailed from another boat, "Seamans! Are you aboard that boat?"

"Yes," I shouted back.

"Mr. Webb wants to talk to you on the telephone!"

We had a low-powered radiotelephone on *Serene*, but there was a good-sized motorsailer in the harbor that had some decent power, so I rowed over there and placed a call to Jim. He said, "We've had Dr. George Mueller in. Both Hugh and I are greatly impressed with him. We feel sure he can do the job, but we want to check once more with you before going ahead."

I knew George as an effective vice president at Thomson-Ramo-Woolridge (TRW), where he had a major role in their satellite hardware developments. "You've got my proxy on that one," I said to Jim. "It would be great if he could join us." On July 23 it was announced that Mueller would succeed Holmes, effective September 1.

In the two years prior to George's appointment, NASA had been highly decentralized, with all of the centers and program offices reporting to me as the associate administrator. Many concerns about this setup had cropped up. Program people said that they, in effect, didn't really have control over their destinies, because they didn't directly control what went on in any center. They had to negotiate with a center director to get things done, in competition with other program offices. For their part, the center directors were concerned that they didn't have anybody in headquarters who was really concerned about their centers as resources.

It was quite true that my schedule didn't permit me to put the needed time on some of the issues. If you added up the number of people on the organization chart reporting to the associate administrator, it came to about twenty—way beyond what all the textbooks on management prescribed. They said that an individual can encompass no more than five or six people reporting to him or her.

When all was said and done, I was very much in favor of the Apollo reorganization that became effective in November 1963. Headquarters staff was reorganized by program objectives—Mercury, Gemini, Apollo, and what became known as Apollo Applications (advanced missions). George Mueller brought in General Samuel L. Phillips to take charge of the Apollo program. He also assigned Charles ("Chuck") Mathews to run Gemini, William Schneider to wrap up the Mercury program, and Edward Z. ("EZ") Gray to take charge of advanced planning. The Apollo and Gemini programs had their own program, systems, test, and quality-assurance groups. These groups had counterparts in the three spaceflight centers.

George Mueller was a determined organizer. He held frequent meetings, maybe monthly, with what he called his board of executives, which included Wernher von Braun from Marshall, Bob Gilruth from Houston, and Kurt Debus from Cape Canaveral. Then a Sam Phillips or a Chuck Matthews would come in and present his program before this group.

George was also the kind of person who worked seven days a week and thought nothing of taking the red-eye overnight. So many charts were presented at his meetings—some showing dollars, some showing schedules, some showing drawings of new designs—that his meetings eventually were referred to as "pasteurized" (one chart after another going *past your eyes*). He was tireless. There was no such thing as Saturday or Sunday to him, and he had his people in all weekend long. There were complaints that George was pushing too hard.

There was another potential drawback to this new organizational setup: the center directors had a lower profile in this reorganization. Because they no longer reported to me, they didn't have the same kind of access to the top of the organization as they had had. I think Jim, Hugh, and I were sensitive to this and tried to find proper avenues for keeping in touch with them.

President Kennedy

I was fortunate to meet and work with President Kennedy on a number of occasions during his thousand days in office. Several were memorable. On July 21, 1961, two days after Gus Grissom's Mercury flight, the President signed our 1962 budget authorization. Hugh Dryden was on vacation and Jim Webb was at the Cape presenting suitable honors to Grissom, so I represented the agency at the signing.

The President came into the Fish Room at the White House surrounded by a great battery of TV cameras. There was a terrible glare of lights. He came over to a table where maybe thirty of us were standing, including Vice President Johnson. He sat down at the table, faced the cameras, and discussed the space program for four or five minutes in a fluent, knowledgeable way, without referring to what looked like prepared remarks rolled up in his hand. He then started signing the bill. He had about three dozen pens there, and somehow or other he managed to use them all in signing, dating, and underlining his signature. He chuckled away as he did so. Then he stood up with a fistful of pens in his left hand and started handing them out. He came to me and said, "Here is one for you."

I said, "Thank you very much, Mr. President. I will give this to Mr. Webb, who is down at the Cape today."

He turned away and gave pens to others, including Senator Kerr. A minute or so later, he was leaving the room when he turned around (he was very lithe in the way he carried himself around a room) and approached me again, saying, "Well, here's one for you, too."

In September 1962, President Kennedy toured NASA's space centers. For this trip, I flew in the President's plane with Secretary of Defense Robert McNamara, press secretary Pierre Salinger, and others. The Vice President took another plane, since the President and Vice President never fly together. Jim Webb and others flew with Johnson.

We stopped first at Huntsville, then went down to the Cape. Then we flew across to Houston. Jim Webb had asked me, since we couldn't stop in New Orleans, to explain to the President what we were doing at the Michoud Assembly Facility in Louisiana, and I invited the President to discuss this on the flight from the Cape to Houston. He said we could talk as soon as we were airborne.

After takeoff, I was sitting facing budget director Dave Bell with my back to Kennedy's compartment. All of a sudden I saw Dave stand up and nearly drive his head through the roof of the plane. I turned around to greet the President, who had come out of his compartment. He had taken off his tie and had on a sport shirt. He sat down with Dave Bell, Harold Brown (director of Defense Research and Engineering), George Miller (chairman of the House Space Committee), and myself.

We spent well over an hour together. He began by listening politely to what I had to say about Michoud, but soon he was entering into the discussion, especially with questions about the military's use of rockets. ("Do they really need the Titan III?") At the end of the discussion, he looked at me and said, "How long have you been with the government?" I told him. "What did you do before then?" I told him that too. Finally I couldn't resist saying that, as a matter of fact, I was a member of his Harvard class. "Gee," he said, "I thought you looked familiar!" For the rest of the trip, whenever we happened to be standing side by side, he would ask informally, "Let's see now, what house were you in?" and "Did you play JV football?" and "Did you know so and so?" He kept at this the whole trip.

In Houston, the President visited the Manned Spacecraft Center, then in temporary quarters, and made a major address at Rice Stadium on the hottest day I can remember. All except the President drove to the stadium in air-conditioned vehicles. Kennedy insisted on sitting in an open car as we drove past tall buildings close to the road. When we arrived at the stadium, we found that the large crowd was in the shade, but that the presidential party was to be seated out on the grid-iron on a temporary stand facing the sun. As I sat there and sweltered, I tried with my imagination to nudge the very few fleecy little clouds over the face of the sun for a moment's relief. Yet the President stood up—completely composed, looking very trim, and not showing the heat at all—and delivered a thirty-minute speech with great enthusiasm. Then he walked out of the stadium, got back into the open car, and drove back down the same street. It was a fantastic display.

We flew back to Washington via St. Louis, where he stopped at the McDonnell plant. At each of the centers, the President obviously enjoyed the opportunity of shaking many more hands than the Secret

Service liked. At each stop, he also gave an informal but very exciting address. These visits were a great boost to morale at the centers.

One evening in mid-November 1963, just after arriving home, I received a call telling me that President Kennedy was thinking of a trip to Cape Canaveral. (The White House is always careful never to be too explicit about a presidential trip until things are all mapped out, for obvious reasons.) Following this, I received a call from Major General Chester V. Clifton, military aide to the President, who gave me more detail. He said that the President wanted to get a feel for how we were coming and that he would have about two hours. What did we recommend?

Julian Scheer, NASA's public affairs officer, came down to my office, and several of us sketched a map on the blackboard indicating where the President might land, what he might see up close, and what he might fly over. We felt that he couldn't cover it all without the use of a helicopter because a couple of bridges connecting the Cape with Merritt Island were not then in use. I called General Clifton back, and there ensued a series of phone calls and discussions of other opportunities for the President while at the Cape, among them a review of Polaris.

The next morning, November 16, 1963, the President flew from Palm Beach to the Cape, where he was greeted by Major General Leighton I. "Lee" Davis and Dr. Debus, the respective heads of military and NASA operations at the Cape, as well as by Jim Webb and myself. He was accompanied by Senator George A. Smathers of Florida, a good friend of his. The President said, with a smile, that he was very appreciative that we had all found the time to be there. Then he stepped into an open car with Jim and General Davis. They drove by the various complexes rather slowly. We joined them inside the blockhouse at Complex 37, from which the Saturn was soon to be launched. There was about a fifteen-minute briefing there with all kinds of models. The President seemed quite interested in what George Mueller had to say. When the briefing was over, he stood up and went over to the models. He expressed amazement at the fact that the models were all to the same scale, because the Mercury launch vehicle was completely dwarfed by the Saturn V. This may have been the first time he fully realized the dimensions of future NASA projects.

We then went out to the pad where the Saturn SA-5 (the fifth Saturn I) was sitting. We stood out in the open with Dr. von Braun, discussing the Saturn and its dimensions. Before leaving, President Kennedy wanted to walk over and stand right underneath the Saturn. This eventuality had come up for discussion with the Secret Service the previous day. They hadn't wanted him to get too close to the rocket. But no matter what anybody thought, President Kennedy was going to go and stand under the Saturn. "Now," he said, "this will be the largest payload that man has ever put into orbit? Is that right?"

"Yes," we said, "that's right."

He said, "That is very, very significant."

We then climbed into the President's helicopter, which had been flown down from Washington on a transport plane for his use. My job was to sit with him as we flew over the new construction area on Merritt Island and to point out the future location of such things as the Vehicle Assembly Building and the launch pad (Complex 39). Afterwards we flew about fifty miles offshore to watch a live test of a Polaris missile. Admiral I. J. Gallatin, who was in charge of the Polaris program, described to the President what he was about to see, which led to a discussion of the whole concept of nuclear submarines—a classified matter about which the President was clearly interested and knowledgeable. We landed on the deck of a waiting ship. The President hopped out vigorously. In honor of his visit, he was presented with a Navy jacket, which, as a naval hero of World War II, he happily put on. He was obviously enjoying himself.

Then, as planned, President Kennedy gave the order to fire. There was a countdown. . .then a hold! I could feel the tension in the Navy personnel there, and I also noticed a couple of Air Force and Army men winking at each other. The President stood watching with binoculars. Fortunately, another Polaris missile was on hand, and the launch was shifted to the backup. When the missile breached the water, we could see that it had "Beat Army!" painted on its side. We got back in the helicopter, and the President wore his Navy jacket for the rest of the trip.

On the way back, he brought up the matter of the Saturn SA-5. "Now, I'm not sure I have the facts straight on this," he said. "Will you tell me about it again?"

I explained (among other things) that the usable payload was 19,000 pounds, but that we actually would have 38,000 pounds in orbit.

"What is the Soviet capability?," President Kennedy asked. I told him. "That's very important," he said. "Now, be sure that the press really understands this, and, in particular, see..." (he mentioned one reporter by name). Just before we landed, he called in General Clifton, his military aide, and said, "Will you be sure that Dr. Seamans has a chance to explain to..." (he mentioned the reporter's name again).

We got off the helicopter and walked quickly over to the President's plane. He shook hands with Jim and the others, then turned back to me and said, "Now, you won't forget to do this, will you?" I said I would be sure to talk to the reporter.

"In addition," he said, "I wish you'd get on the press plane that we have down here and tell the reporters there about the payload."

"Yes, sir," I answered. "I'll do that."

Six days later, on Friday, November 22, I was holding a meeting in my office, when I got a call from Nina Scrivener, Jim Webb's secretary. She said, "Something dreadful has happened in Dallas. You'd better come on up to Jim's office."

"You mean the President's been hurt?"

She said, "It may be worse than that."

I closed down the meeting very quickly, then called Gene. "Watch the news," I said. "I don't think we're going to be having that NASA gathering tonight."

Jim Webb had three televisions in his office, so that he could have all three networks going at once and flip on the sound of the one he wanted to hear. We sat there watching all three networks. Finally, Walter Cronkite came on and said the President had died. Gene arrived at the office a little later, to distribute the food we had planned to serve at dinner.

The following day, we had scheduled our regular monthly program review. I argued strenuously that we ought to go ahead with the meeting, that President Kennedy had been very interested in NASA's programs, and that he would have wanted us to press ahead. As a result, we were probably the only federal government organization doing business that day. When the review was over, Jim turned to Hugh and said, "I'm going over to the White House. Do you and Bob want to come along?"

The three of us and our spouses stood in line in the East Room, where the President was lying in state with a Marine at attention by his side. Everyone wore black. The casket was closed and draped with the flag. There was immense grief on every face, and many significant symbols, such as the Great Seal of the United States of America over the door, were draped in black. There were no flowers, no music, only the murmur of hushed voices and the shuffle of feet. It was the saddest place and the saddest time in our lives.

The following day I was scheduled to give a talk at church. I had to comment on the assassination and tried to be positive, not maudlin. On Monday, Gene and I walked over from our Georgetown house to watch the funeral procession, pushing through the mourning crowds outside Arlington National Cemetery. After the committal, fifty fighter planes, representing the fifty states, flew over our heads in formation. They were followed by Air Force One, the President's own. The same plane that had flown President Kennedy to Cape Canaveral nine days before dipped its wing as it flew over his final resting place.

President Kennedy's assassination had a profound impact on the peoples of the world, and particularly on those working closely with him in the government. Those responsible for launching the Saturn SA-5, which he had observed and commented on during his inspection in November, wanted some way to express their gratitude for his interest and their grief for his loss. Rumors were rampant that special markings would be placed on the Saturn, which led to the implementation of special security provisions. In the aftermath of the launching, while still in the blockhouse, we all felt such an emotional upswelling that there was a near-unanimous request for a call to Mrs. Kennedy. I felt, perhaps wrongly, that such a call would be upsetting for her, and I suggested instead that I carry this thought back to her.

When I returned to Washington, I contacted Walter Sohier, NASA's general counsel and a friend of the Kennedys. He didn't think Mrs. Kennedy would be interested in a visit, so imagine his surprise when he and I were invited for tea the following afternoon! Mrs. Kennedy was very gracious; sat patiently as I explained the circumstances of our being there; brought in her children (both recovering from chicken pox); and sent us away exhilarated by our encounter.

My letter to her of February 7, 1964, which follows below, is self-explanatory, but her response of March 14 was unexpected and deserving of comment. Mrs. Kennedy had had little time to move out of the White House into a house on N Street, Georgetown, loaned to her by a friend. Remarkably, she started immediately to reply to the huge number of people who had attended the funeral service or offered their condolences in other ways. Gene was among many who volunteered their assistance. Hence her longhand response to my visit and letter is truly remarkable.

February 7, 1964

Mrs. John F. Kennedy
Washington, D.C.

Dear Mrs. Kennedy:

Thank you for the pleasant visit you afforded us Monday evening. It meant a great deal to me to be able to tell you about the recent Saturn launch from the John F. Kennedy Space Center.

The accompanying detailed engineering model of the actual Saturn launched on January 29th is presented to you with appreciation from all of us. It was utilized by Dr. Wernher von Braun and the staff of the George C. Marshall Space Flight Center which has responsibility for the Saturn development.

Having seen young John's interest in space toys, and having barely escaped from your home with the other model that I brought, I am also sending some fairly sturdy launch vehicle models for his enjoyment.

Sincerely,

Robert C. Seamans, Jr.
Assoc. Administrator

MRS. JOHN F. KENNEDY

March 14, 1964

Dear Dr. Seamans—

I do thank you for that most precious model of the Saturn—the one that Wernher von Braun and everyone worked on—(I could not believe my eyes when Walter Sohier brought it.)

John had a fleeting happy look at it—and then I sent it to Archives—to go in Jack's library.

Your thoughtfulness has touched me so much—that you would wish to come—and tell me about the Saturn booster—and think of calling me from the blockhouse when it was going off. All I care about is that people still remember what Jack did—and you were always thinking of him.

Then when you came and saw John—it was so kind of you to see how a little boy who had grown up so close to a father who always had exciting new plane and rocket models in his office to show him—who took him on his most cherished plane and helicopter rides—would still care so much about all those things—and feel so cut off now that they are no longer a part of his life.

Those "heavy duty" models that you sent him are his joy—taken apart and put together constantly—I do thank you more than I can say, for your thoughtfulness to him and to me—

Sincerely,

Jacqueline Kennedy

Family Times

The great tragedy of Kennedy's assassination notwithstanding, Washington continued to be pretty exciting, not just for me but for our entire family. Just after the first successful Gemini flight in March 1965, I got a call from Jim Webb's exec saying that I, together with the two astronauts, Gus Grissom and John Young, was to receive the NASA Distinguished Service Medal at the White House. He wanted to know whom I wanted to invite for the ceremony. I said, "Well, obviously I'd like to have Gene and our children there. It would be wonderful if I could have my mother and father there, too, and if the Lorings could be there, so much the better."

He said, "I don't know if Jim is going to go for all of that." Webb was a little prickly about it, but all were invited and all were there, except for Kathy and Lou, whom we couldn't locate in time. A letter Gene wrote to her mother describes the whirl she and the children found themselves a part of:

Dear Mother:

This last week has been wild. Since writing you we've had one experience after another. [Bob's parents] arrived to visit Thursday evening along with Joe and Toby, who returned with Bobby tanned and smiling from Cape Canaveral. . . .

The next morning we all drove in our best to NASA headquarters. Joe wore Congressman Moorehead's shirt, Toby's jacket, Bobby's shoes and socks, and shaved with his Grandfather Seamans' razor. He was a bit scruffy. From NASA headquarters we went whizzing by motorcade behind a police escort through all the red lights to the front door of the White House. The children loved such an exciting ride. Arriving a little early at the White House, we were told by an aide to make ourselves at home. This meant that we could wander at will through the dining room, the Green Room, Blue Room, Red Room, and so on, all so beautifully refurbished by Mrs. Kennedy and filled with glorious bouquets of flowers.

The citation I received that day was a true Webbian creation. He could really turn on the accolades in a most gracious way. It was given to me because I was the general manager of an organization that involved "50,000 government workers and 400,000 contractor employees," but it went on and on about me, the general manager, in such a grandiose way that only Mother and a few others would believe it!

The President and Mrs. Johnson had pictures taken with the astronauts. Out of the blue, Jim said very loudly, "Don't forget Bob Seamans!"

The President said, "Fine," and all of a sudden in a room full of people—and before national television cameras—my whole entourage trooped up beside the President and First Lady, who impressed all of us by their warmth. A few days later, my parents sent us a picture from the *Salem Evening News* of Bob Seamans and his wife and family at the White House, with (according to the caption) astronaut John Young's son in the foreground. My mother's note read, "What a striking resemblance between John Young, Jr. and Daniel Seamans!" The boy in the picture was Daniel, then six years old.

Robert Wagner, the mayor of New York City, was very anxious to have the astronauts feted on his turf whenever we could supply them. He used to say, "You have no idea how few handles we have in our school system to motivate our kids. It's just wonderful to have these young astronauts come to New York and for all of our students to see them." Following the White House ceremony for the Gemini astronauts, Wagner invited Grissom, Young, and their spouses for a ticker-tape parade. Webb said, "Fine, we'll send them up, but Bob Seamans is going to come along, too." Jim Webb felt that the astronauts were getting too much credit for the space program. NASA wasn't just a lot of skilled hot rods up there in space, he said. It was also a huge number of technical, scientific people. He felt the country ought to be recognizing their contributions as much as it did the astronauts'.

"That's fine," Wagner answered. "I'm glad to have Bob Seamans come along."

But Webb went on. "He's not just coming along. He's going to get the same Medal of Freedom and the same key to the city that the astronauts do."

Wagner balked at that. Since Charles Lindbergh had received it, the Medal of Freedom hadn't been given to anyone except Alan Shepard and John Glenn. Now it was being given to two Gemini astronauts—and Bob Seamans?! He finally relented, however.

Vice President Humphrey was our escort, providing enthusiasm and wit, and Gene was able to accompany me. We found ourselves whisked everywhere by motorcycle escorts and the Secret Service. As we drove down FDR Drive, fireboats on the East River were arching huge streams of water while tooting their horns. Later, during the ticker-tape parade, looking up at the skyscrapers with showers of paper blowing down on us was a dizzying experience. Next, we stood in a receiving line at the Waldorf and shook hands with at least 2,000 VIPs. In the course of the day, we met some very interesting people, including Mayor Wagner, U.N. Secretary General U Thant, U.S.

ambassador to the U.N. Adlai Stevenson, and famed newscaster Lowell Thomas.

Despite my endlessly hectic schedule, the Seamanses did have some enjoyable moments away from the government during the 1960s. In 1965, we took a memorable family trip to Europe. Gene planned it for Christmas vacation. We all wanted to ski, but she thought we also ought to do something worthwhile to improve ourselves. So we laid plans to spend three days in London first, before flying on to Switzerland. Everything was going to be wonderful.

Three days before our scheduled departure, Jim Webb told me he was flying down to the Johnson ranch in Texas in a couple of days. He said it was very important that I go with him because I was familiar with the facts in the case.

I said, "I think I told you, Jim, that's the day we're leaving for Europe."

He didn't take that very well. "Well," he said, "your family can go to Europe, and you can join them a couple of days later. Surely that won't be any big problem."

That night I discussed it with Gene. "Something's come up," I said. "It's really important. A decision has to be made down at the ranch, and Jim wants me involved."

Gene said, "If you think I'm going to take Daniel, May, Joe, Toby, Kathy, and all their ski equipment alone into London for three days, maybe we ought to just give up the trip."

"Now, wait a minute," I said. "At worst, we can fly directly to Switzerland and give up London." Gene didn't like that at all. She had put a tremendous amount of effort into finding accommodations and activities in London.

The next morning I went back to Jim and said, "I've talked to Gene, and here's the way it is. Either I go with her and the family, or we're giving up the trip entirely."

"Okay," he relented, "I guess I can handle it alone." It was evident to me, though, that he was very unhappy about having to do so.

Gene and I had planned to fly to Boston from Washington with Kathy, May, and Dan; meet Toby and Joe, who were going to school in the Boston area; and fly over on BOAC from Boston. Lou decided to cross the Atlantic aboard a tanker and meet us in Europe. But by the time those of us from Washington got to Boston, BOAC was on strike. The only thing we could do was take the shuttle to New York for a connecting flight on Alitalia. The weather was bad, and the airports were mobbed with students heading home for the holidays. We couldn't all get on the same Eastern shuttle to La Guardia. I went ahead to make sure that our equipment ended up in one place. The airline said they couldn't guarantee that our luggage would fly with us. There was no simple way to transfer from La Guardia to Kennedy, where the Alitalia flight departed. We had to commandeer a whole fleet of taxis. In the midst of all my running and shouting, a stranger hailed me. He thought I was a redcap. I helped a couple of women with their luggage, and they handed me five dollars, which I refused. When we finally arrived in London, we discovered that London cabs have no luggage racks (for skis). We had to hire one cab just to carry our equipment, with the skis barely loaded inside on a diagonal. Somehow Toby and I got in with them. The children were quite hysterical until they realized that *everyone* in England drove on the left!

In Rome, May had another adventure awaiting her. A year or two before, Jim Webb had received a call from Father Hesburgh, head of Notre Dame University and a personal friend. Hesburgh had said Pope Paul VI was very interested in the American space program and would appreciate receiving special articles and reports. Before leaving for Europe, I got in touch with Father Hesburgh and explained that not only would we be delighted to give him any material he wanted but also at some suitable time, if His Holiness would like it, we would be glad to brief him on what we were doing.

Father Hesburgh thought that was a great idea. It turned out he was going to be in Rome when we were. When we arrived in Rome, we had a message waiting from him. "No news yet," it read. "Haven't been able to arrange anything so far." The following morning we talked on the phone. "You've got to realize," he explained, "that His Holiness has a lot of constituents, and it's not easy to make arrangements. Things aren't always as organized over here as they are at home, but I'm working on it."

Saturday morning he said, "We have an appointment for five o'clock." I asked whether I might bring along Gil Ousley, our NASA representative in Europe. Father Hesburgh said that would be fine. I mentioned that neither Gil nor I was Catholic. He said that didn't matter. At four o'clock Father Hesburgh arrived at our hotel, the Hassler, with another priest who had on a great wide-brimmed hat. "Where are you going while we're at the Vatican?," Hesburgh asked Gene.

She said, "My daughter May and I are going out shopping."

"Oh, come with us," he said. "Why should the men have all the fun?"

So Gene, May, Father Hesburgh, the priest, Gil, and I all jumped into a small Fiat in the pouring rain and went steaming through Rome. As we approached the Vatican, Father Hesburgh said, "I always keep right on driving until somebody stops me." We went through a gate off St. Peter's Square, right past an officer of the Swiss Guard. We finally arrived in a courtyard and took an elevator with a St. Christopher Medal mounted on the wall up to the Pope's apartments, where we were met by one of his assistants. Gene and May were taken to sit in the Pope's private chapel, where May was especially intrigued by the reliquaries. Gil and I were taken to a small meeting room where Gil set up a projector. We had a filmstrip of photos taken by the Ranger as it approached the Moon. I had also brought along a large album filled with lunar pictures—all carefully laid out and documented—for the Pope to keep.

At about two minutes before five, I was standing by the door through which the Pope was going to come. I asked Father Hesburgh what to do when the Pope entered. He said, "First I'll kneel down and kiss his ring. Then I'll stand up and introduce you. Just shake hands with him."

"Okay. At the end, how will I know when to stop?"

"You'll know."

In came Pope Paul VI—a surprisingly little person, no more than five feet, six inches tall. When he sat in his gilt throne, his feet didn't touch the floor. They rested on a footstool. I said what a wonderful opportunity it was to have this chance to talk about the space program with him. I presented him the book, along with a letter from President Johnson, then said I thought he might like to see a few pictures from the Ranger mission. I showed them and made a few more remarks about NASA. The whole meeting couldn't have taken more than fifteen minutes. Then the Pope stood up, came towards me, shook my hand, and thanked me (speaking all the time through an interpreter). He presented me with medallions for Jim Webb and myself. "I understand that your wife and daughter are here with you in Rome," he said. "Where are they now?"

They were waiting in the chapel.

"I want to meet them," the Pope said. He was politely informed by an assistant that perhaps May in particular was not dressed for an audience. She still had on the yellow sweater and rain boots she had planned to wear shopping. "That doesn't matter," the Pope said.

As Gene later told the story, she and May were quite relaxed and enjoying the quiet of the Vatican when a Swiss Guard came running down the corridor saying, "*Veni, veni*!" They knew something was up. May, after following him along dark corridors, came tearing into the room from behind a tapestry, almost out of breath. She then composed herself and did a very nice curtsey—taught to her, no doubt, by the National Cathedral School for Girls. Gene followed and shook the Pope's hand, and His Holiness presented rosaries to her and to May. Then Pope Paul very kindly said good-bye and left with Father Hesburgh for another meeting, leaving Gil and the three Seamanses in the hands of the priest with the big hat.

In the down elevator, we were elated. It had been an amazing experience. Riding with us in the elevator was a wealthy Italian couple, dressed to the nines and steaming mad. They started gesticulating and almost yelling at our guide. He responded in a mild sort of way. We didn't know exactly what was going on. Finally in the car Gil asked him, "What was that all about?"

All he would say is, "Some people think they know more than the Pope." Apparently, the Italian couple had expected a private audience and didn't get it. By our demeanor, they could tell we had, and they obviously didn't like being shoved aside by Americans who hadn't even dressed for the occasion!

More Travel Overseas

Two years later, Gene and I left the country with a complicated itinerary. We first went to Paris so that I could meet with a NATO group known as AGARD (Advisory Group for Aerospace Research and Development). Each country sent three national delegates, and I was one of the three Americans.

From there we went on to Kenya, where I wanted to visit the San Marcos Project, a joint venture between the United States and Italy. We provided the rockets; the Italians built two platforms near Rome, one for launch and the other for control and observation. These were towed through the Suez Canal to the coast of Kenya, close to the equator. Professor Emilio Broglio was in charge of the project. Gene and I flew to Rome to join him for connecting flights to Nairobi and then Mombasa. At the Rome airport there were again TV cables all over the place, but not because of our visit. It turned out that Josef Stalin's daughter, Svetlana, who had recently defected from the USSR, had arrived a couple of hours before.

Mombasa is a fascinating seaport, and we had very nice accommodations there in an ancient house-hotel overlooking the city. On March 11, we celebrated Gene's forty-fifth birthday with a joyful dinner hosted by the professor. The next day we took an amazing, bumpy drive in an oversized jeep with a Turkish driver. We drove up the coast through places where the brush was strewn everywhere from elephants having run amok. We finally reached the project base camp in a place called Campo Basa just north of a little town called Melinde. The camp had grass huts, and there were lots of cute monkeys swinging from the trees. We boarded rubber boats with outboard motors and headed out to the two platforms. To get us aboard, a crane was lowered to lift our entire boat from the water. Here at the equator, seated beneath a canvas canopy on a very warm day, we had one of the best Italian meals we've ever had. A Kenyan steward provided the toast, welcoming the Americans and thanking them for their contributions. He was a fierce man, who may have been primed with a bit of brandy. He shouted his remarks and ended with a startling yell, brandishing a machete.

The next day it was back to Nairobi for a flight to New Delhi. There we were hosted by the Galen Stones, a couple from the Boston area whom we knew in Washington. He was the number-two person at the American embassy. In fact, he was the one Svetlana had declared herself to when she had defected. So it was fun for Gene and me to hear firsthand how it had happened. We returned home via Australia, where Prime Minister Holt and I dedicated an Apollo tracking station at Honeysuckle Creek.

LBJ

Once when Jim Webb was away, I received a call saying there would be an important meeting with President Johnson and the heads of all federal departments and agencies in the Cabinet Room. Using barnyard vernacular, the President gave us quite a tongue-lashing about the need to cut expenditures and asked for a 5-percent cut across the board. When he was done, he said that he expected each of us to shake his hand as we filed out, to look him in the eye, and to say we would reduce the spending in our respective agencies by 5 percent.

I had time to agonize over this, because we left the room in order of rank, beginning with the secretary of state. I was just ahead of General William F. ("Bozo") McKee, head of the FAA. When I reached the President, I said, "Of course, we'll do everything we can to make the reductions you've asked for, but I'm sure you wouldn't want to jeopardize the lives of the astronauts in so doing."

The President didn't like that at all. Bozo caught up to me after both of us had left the White House and said, "You ought to know that the President said to me, 'Will you take that young guy Seamans out behind the barn and give him the facts of life?'"

President Johnson usually knew what he wanted and made no bones about asking for it. One Friday morning in August 1965, I was in the shower when Gene came running. "Bobby, come quick! The President's on the phone."

"Good morning, Bob," Johnson said, just as sweet as could be. "I just wondered if you'd care to come over to see me on my birthday today."

"Yes, sir. Happy birthday, sir." Jim Webb was out of town, and it was clear to me that I was invited in his place, as representative of the Office of the Administrator.

When I arrived at the White House, I was told the President was busy; so I went back to my office and returned later. My second time at the White House, I waited and waited and finally got in to see him. He pretended not to notice me, then suddenly turned and said, "Seamans, there you are." His tone had clearly changed since the morning phone call. "Sit down! Seamans, you guys over there at NASA, you're pretty good with your science and you're pretty good up there on the Hill, but as far as I'm concerned, Seamans, you're a great big zero. You know why you're a great big zero, Seamans?"

Before I could answer, he asked, "What do you think is the most important thing in my life, Seamans?" Answering his own question, the President went on, "It's peace. It's peace." Suddenly I knew what the problem was. He and Jim were always disagreeing over whether astronauts should be used as national emissaries. Jim fought him like a steer, saying he didn't think it did any good. We had astronauts Cooper and Conrad in orbit aboard Gemini 5 at the time. It turned out that, upon their return, the President wanted to send them on a peace mission.

Then he said, "All right, Seamans. I want Jim Webb to be with me this Sunday down at my ranch in time to go to church with Dean and me." "Dean" was Dean Rusk, Johnson's secretary of state.

I said, "Mr. President, Jim Webb has been working very hard lately and has been very tired. He's hiking with his family in the mountains of North Carolina."

The President retorted, "Jim Webb is the best goddamn administrator we have in this government. You call me in one hour and tell me he's going to be there."

I went speeding out of the White House, got to my office, and fortunately reached Jim in Asheville, North Carolina.

Jim's response was: "You tell the President of course I'll be there. But Bob, since you obviously have the flavor of this thing, I think it would be awfully nice if you could be there too. You ask the President if you can come."

I called the White House and got through to Johnson immediately. "Mr. Webb would be very happy to be with you," I said. "He wonders if I might come along, too."

"Why, Bob, of course, come along! Bring your wife! Bring your family!"

I said, "That won't be necessary, sir. If you need to reach Jim and me, we will be at the Manned Spacecraft Center in Houston."

"Oh, Bob, don't do that. You and Jim come spend the night with me at the ranch."

I said, "I'll check with Mr. Webb, and we'll advise your office."

I called Jim Webb back to tell him the President did want me to come and also wanted us to stay with him at his ranch. I asked Jim what he was going to wear. "You hear about people going out on horseback at the ranch," I said.

"I'm going to wear business clothes. You do the same. No fooling around. We're going to go in there, do our business, and get out." It was good advice.

We ended up spending Saturday night at the LBJ ranch in a small house separate from the President's quarters. We arrived late Saturday afternoon and waited in the President's study while he attended a neighbor's barbecue. All of a sudden we heard the slap, slap, slap of the rotors of the presidential helicopter. Pretty soon the President came in, accompanied by Secretary of State Dean Rusk and Bill Moyers, Johnson's public affairs officer.

"Okay," the President said, "I've got the press release about the astronauts' goodwill tour here, but I want to get your comments on it, Jim. We'll let Bill read it."

Moyers started in, "After the most successful space flight of all time," and Jim cut right in.

"It wasn't that successful, Mr. President," he said. "We had trouble with the fuel cell."

"All right, Jim," Johnson said. "After the least successful space flight of all time—" While Jim bristled, the President gave me a big wink.

When the release had been read, Jim put up an argument, though not too strenuously.

"Okay, Jim," the President said. "If you want to read it over carefully and comment, you can look over Bill's shoulder, and if he thinks it's appropriate he will make the change." Johnson wasn't about to let Jim Webb hold onto the piece of paper!

"No, sir," Jim said, "it's fine as written."

The following morning, the astronauts landed at about nine o'clock. Johnson had a phone hookup to the ship that retrieved them and spoke with them, while the press gathered around him. He told them that they were going on a goodwill mission for the country. They, of course, said, "Yes, sir!"

We went back inside the house, where all the venetian blinds were closed so that the press couldn't snoop. Johnson was a very secretive person, and he didn't want anyone to know that with him in the house, in addition to Dean Rusk, were Larry O'Brien, the postmaster general-designate, and Arthur Goldberg, the Supreme Court justice, whom

Johnson was going to nominate ambassador to the United Nations. At one point he even said, "Get away from the windows!"

Then he organized the trip to church. "We leave at quarter past ten. Dean and Jim, you're going to ride with me. Bob, you go with Bird [Lady Bird Johnson]," and so on.

We headed off toward church as scheduled, with the President and First Lady driving their respective cars. We stopped at the cemetery so the President could put a wreath on his grandmother's grave, then visited a replica of the tiny house where Johnson was born, built on the site of the original house. We all trooped inside to look at his booties and other childhood memorabilia. We headed into Johnson City, where he wanted to show us the house where he and Mrs. Johnson lived when they were first married. He stopped in front of it, got out, and flagged down the First Lady's car.

"We're going to be late for church!," she protested.

"Goddamn it, Bird, get out of the car and go through the house!" So we all got out of our respective cars, tore through the house, and jumped in the car again. When we reached the very simple church, there were four or five Greyhound buses for the press outside with their engines throbbing. I helped Mrs. Johnson out of the car, then stepped aside as the President approached. I naturally thought that he would accompany his wife into the church. Instead, Johnson said to me, "You take Bird in." So I marched up the church steps with the First Lady, ahead of the President and secretary of state. Lady Bird introduced me to the minister, who was waiting for us on the doorstep. Then we started down the main aisle of the church, which was packed except for the front pew on the right. I figured that if I followed her into the pew, I would end sitting between her and her husband, so I volunteered to go around to the other side. The pew quickly became very crowded with the President, the First Lady, both of their daughters, the secretary of state, Jim Webb, and so on. I whispered to Mrs. Johnson that I would slip to the back of the church. The President heard me. He turned to the people in the pew behind us and said in a very loud voice, "Will some of you back there please move so Dr. Seamans can have a seat?"

As we were driving back to the ranch, Mrs. Johnson said, "There's going to be another press conference, isn't there, when we get back to the ranch?" Nobody answered. Everybody knew, but one could get into

trouble talking about something the President was to be involved in—
Johnson's secretiveness again. Colonel Emmerson ("Mike") Cook, who
worked for me years later in the Air Force, told me that sometimes
Johnson wouldn't even tell his pilot where he was flying. "But, Mr.
President, I can't take off until I know our destination."

"You take off and head towards Kansas City! When we're airborne I'll tell you where we're going."

Mrs. Johnson repeated her question. "Isn't it true there's going to
be a press conference?"

I finally spoke up. "I believe there is. I think I heard somebody
mention something about it."

"Well, what time is it going to be?" she asked.

I knew what time it was scheduled for, and so did the others in the
car, but nobody said anything. Finally someone said, "We're not quite
sure, Mrs. Johnson. Things might have changed."

Driving along the highway at about sixty-five miles per hour, Mrs.
Johnson picked up the radiotelephone on her dashboard and said,
"Can anybody tell me where the President is?"

A voice: "Mrs. Johnson, he's two minutes and ten seconds ahead
of you on the highway."

"Can you tell me what time the press conference is called for?"

"Twelve-thirty."

Then she said, "There's still birthday cake, is there not, from the
President's party yesterday? I want to have it out for the press on the
table under the oak tree. The pieces were too large yesterday, so have
John at the ranch cut them all in two."

"Yes, Mrs. Johnson."

Sure enough, when we arrived, the press was there ahead of us and
so was the cake! We had a very nice lunch, after which Jim and I
departed. On the way back on the airplane I said to Jim, "I don't know
Mrs. Johnson as well as you do, but I think she's something else."

"If there's a saint on Earth," he said, "she's it."

End of the Triad

NASA sustained a terrible loss when Hugh Dryden died of cancer on
December 24, 1965. Hugh had played an extremely important role at
the agency. He would have been the first to say that he wasn't an

administrator type, one who could hammer on Congress one day and the President the next, then lead a charge out to the Midwest to take on an industrial delegation the next. Nor was he the person to keep the pressure on programs morning, noon, and night, in order to keep projects on schedule and within funding limits.

What he was, rather, was a man with excellent judgment on what NASA's objectives ought to be. He understood how things got done—and snarled up—in the government. He was also effective in relations with the university community and highly respected in the international community. Finally, he was a balancing agent for Jim Webb, on the one hand, and me, on the other. Both Jim and I frequently needed guidance. Together the three of us formed a triad. I used to meet with Hugh and/or Jim at least once a week with a list of up to two dozen items that I felt we had to agree on. We normally clicked through them, rapid-fire. That was the way we operated for five years. We disagreed on a few occasions, but on the big issues, we almost always came to unanimous agreement before taking action.

Hugh never let on that he was seriously bothered by his cancerous condition. But by 1965 he was undergoing chemotherapy that would have been pretty difficult for any human being to take, even one who wasn't reasonably active. He showed tremendous courage. While he was at the National Institutes of Health for treatments, his secretary, Jo DiBella, would go in every day to give him papers to sign. Jim and I would also go in and discuss issues with him. A few days later, Hugh would come out of the hospital and often fly off to Japan or some other distant place.

Still, we were obviously approaching the time when we were going to have to make some changes because we were not going to have anything like his full-time services. One night not long before Hugh died, Jim drove me home from a function in his trademark black Checker Cab. (Although he still used a driver who was separated from the back seat by a window, Jim did not want to be accused of using a limousine. "It's the little things that can get you into trouble in Washington," he noted.) When we arrived at Dumbarton Rock Court, he stopped me from getting out. He told me his plan was to submit my name to the President as the next deputy administrator of NASA. I thought that perhaps it would be better if he didn't. I felt it was highly desirable to have both a general manager and a person like Hugh Dryden as deputy. Shrinking the triumvirate to a twosome seemed to me unwise.

Furthermore, I liked my job. I thought I understood it, and I felt that I could continue to be effective in it. I thought we had come a long way down the pike on the projects that had been initiated since I had first joined NASA, and I believed that to stay with those projects would be a logical way to wind up my work in the government. My feeling was that going to the Moon was a unique venture and that if we had an organization that was working, we shouldn't mess with it. We could all stand another couple of years of long hours, so why monkey around with the assignment? In addition, I was by no means sure I could be an effective deputy administrator. I wasn't a member of the National Academy of Sciences and couldn't perform a lot of the roles Hugh had taken on. So I pushed hard to stay where I was and to bring in a new deputy for Jim.

We finally came to the sad moment when Hugh wasn't with us. Jim told me that soon afterwards, when he was with President Johnson, the President asked, "What are you going to do about your vacancy? I suppose you're going to have Bob Seamans become the deputy?" Jim acknowledged that he was, and that was that. I immediately was given an interim appointment because the Senate was out of session. At my Senate confirmation hearing some time later, Senator Edward M. Kennedy said nice words about my service in his brother's administration, and Senator Leverett Saltonstall said I must be all right because my mother-in-law had been his wife's schoolmate. Following the hearing, I was formally confirmed by the Senate.

"Now, this is going to be a lot harder job, you know, than being the associate administrator," Jim said to me. "You've got to think in terms of the total responsibilities of the government. You can't just think in terms of projects. Other people have got to do that. You've got to work closely with me on the policy issues."

I moved from the sixth floor in Federal Office Building No. 6 to the seventh, where Jim's office was located. I initially had an office where Hugh's had been, in a corner opposite Jim's. I would have preferred to keep this distance, but eventually Jim insisted that I move into an office practically adjoining his. Jim had had a very close relationship with Hugh. Now that I was his deputy, I imagine he expected to have a similarly close relationship with me. I had seen a great deal of him before this, but usually on my terms, when I had had a problem or when we had had full meetings. Now, with his office next to mine, we were cheek to jowl. I began to realize that I was feeling some of the pressures from Jim that Hugh had absorbed previously.

Not too long after I became the deputy administrator, Jim began to grow impatient with the organization. He said we had a system of management that was much too reliant on the two of us and our special capabilities and that he wanted me to help him build an organization that would provide continuity of management after he and I left. We were both working to our limits, he said. Could we come up with an organizational pattern that would not only spread the load in a more reasonable way but also permit us to do a better job and to have better information available for everybody? Jim liked the concept of the secretariat, which he had worked with while undersecretary of state. The secretariat, as Jim conceived of it, would be a central clearinghouse for communications within NASA. All written materials would go to the secretariat, which in turn would decide what level of management should receive it. Things would be passed up and down the chain in such a way that there would be less need for direct, one-on-one communication. I viewed this idea with some skepticism. Never in my technical experience had I seen information successfully transmitted through people who didn't know much about the subject.

Given a free hand at reorganizing NASA, I would have had Earl Hilburn, who had been one of my deputies, continue wading into the really tough technical and management issues. Willis Shapley had joined us in September, three months before Hugh's death, and had worked with him as a chief of staff for external affairs. I thought it would be best to have him continue this work, so that, in effect, Jim and I would have available to us two senior people for help with internal and external affairs, respectively.

Jim did not like this approach. "If you're not around," he told me, "I want to have one person I can go to who knows what's going on. I want a chief of staff across the board, and I want Shapley in that job. To give him something to button onto, I think he ought to take on the secretariat."

I said, "Well, this is going to be quite a load for him to take on.

The concept of the secretariat is not fully understood and accepted throughout NASA." All this did was upset Jim. I had a lot of confidence in Shap, and I was certainly very happy to work with him. He had been the senior person in the Bureau of the Budget examining the programs of NASA and the Department of Defense. I was concerned, though, that I was going to continue with all the responsibilities I had had and in a somewhat weakened position. I would not have as close a tie with Earl Hilburn, whom I predicted would leave under this arrangement, which he did.

During this period, Jim also put a lot of emphasis on job descriptions. He gave Jack Young, the deputy associate administrator for administration, the job of drawing up the entire organization in legalese. We had never done this under the old regime, so it involved a lot of new work. Jim wanted a clear-cut outline of what every person's job was and with whom that person dealt. My mind tends to work differently. I got lost trying to decipher an encyclopedic listing of job descriptions and, from this, to decipher how all the pieces fit together. I liked to look at a line chart and see where the responsibilities were. This told me if something had to go all the way through to the administrator for a final decision or if it could be decided at another level. There are different categories of decisions, which such a chart makes clear.

"Jim, you're absolutely right," I told him. "We can leave a better organization behind us when we exit NASA. But why at this critical juncture in the Apollo program do you want to change it?" I ended believing that he wanted to leave his stamp on the organization and that this was his way of doing so.

The Jack Young exercise did not go well. Jack ended up with stacks of paper, which I was asked to edit—a great big diversion at a time when we were coming down to the short strokes on going to the Moon. No job descriptions were approved, and there came a day when Jack elected to leave NASA.

In November 1966, Willis Shapley and I were invited to Jim Webb's house to discuss possible organizational changes—though *invited* may be too mild a word. Shap and I presented our ideas to Jim, who obviously had a lot of things on his mind. He didn't seem to understand, much less accept, our views. The whole next week Jim was tied up giving a series of management lectures at Princeton. While he was away, a memo from him came across my desk. It said he was very troubled about my views on organization because he was not sure that I fully comprehended the problems of running a major organization in the government. He asked me to put my views in writing.

I thought at great length about the real issue facing us, then made a date to have lunch with Jim as soon as he returned to Washington. On Saturday we went to the Metropolitan Club, and I went through my views on what I thought should be done. He said this was very helpful and fine, but "When are you going to get me that paper?" So I was still on the hook for a written summary, which I delivered to him the following week. To my mind, this was just another diversion from our major responsibilities.

In my paper, I recommended new roles for several individuals. On the basis of my recommendations, we reassigned these and others, and we set up a task force to review what needed to be done organizationally. The task force did not come in with a plan during December or January. On January 27 at 6:31 p.m., a catastrophe occurred that relegated our organizational discussions to second place.

The Fire

NASA had had other accidents in its short history. For example, we had lost our top pilot, Joe Walker, at the Dryden Flight Test Center in the California desert. He was flying his test plane wing-on-wing with a B-70 and was swept up underneath the bigger craft. We also had lost two of our astronauts when they were coming in to land at McDonnell's facility in St. Louis. They had come down out of a low overcast and were not lined up with the landing strip. Instead of going back up and using normal procedures, they had stayed under the overcast, circled around for a landing, and smashed into a building.

We also had had close calls on a couple of missions. On Gemini 8, with Neil Armstrong and Dave Scott aboard, we tried to do something never done before—to rendezvous and dock with an Agena rocket that was already orbiting, then to have the docked vehicles carry out some maneuvers. Everything went well. The docking worked. Then, over China, the two vehicles started to spin.

That evening, March 16, 1966, I was at a large ceremonial dinner at which Vice President Humphrey was the principal speaker. At the beginning of dinner, I had made a special announcement that we had a situation developing in orbit that might have a devastating result. People didn't believe it at first. When they realized I was serious, the whole tenor of the dinner became electrified. I was getting word from headquarters as developments unfolded, but I didn't have very complete information. Nevertheless, people expected me to know. Someone handed me a phone. It was Walter Cronkite asking whether we might lose the astronauts. After dinner the Vice President got up to speak. We had agreed beforehand that, if I got word that the astronauts were safe, I would let him know and he would convey the news. Our hope was that he would be able to announce a happy result before the end of his speech. Humphrey could talk for a long period of time, but even he was starting to run out of words when we finally received encouraging news. He was able to say that, while recovery hadn't yet taken place, Armstrong and Scott appeared to be safe.

Jim Webb was not very happy with my performance on this occasion. He felt that I had taken too big a gamble by publicly announcing information when I didn't have all the facts in hand. He was right, and it was obvious that we needed to do something a little different in the future, if and when we did have a mishap. I went back and looked at the procedural document for accidents and brought it up to date. I also made up my mind that if such a situation arose again, I was going to get to my office or someplace where I had good communication immediately. I felt that this would allow me to deal more effectively with situations that had serious implications and in which the media had intense interest.

Still, neither I nor anyone else at NASA was prepared for the Apollo 204 fire of January 27, 1967. The timing of the accident—in which astronauts Gus Grissom, Ed White, and Roger Chaffee suffocated to death from a fire that broke out in their capsule during a test on the pad—was incredible. It occurred the day President Johnson signed a space treaty with the Soviet Union at an elaborate White House affair. Jim Webb used the occasion to honor the contractors who had been working on Gemini and to make known to the contractors' senior people on the Apollo program the importance of what we were doing in the international arena. The ceremony was to be followed by a formal dinner, to which the top executives of the space program were invited.

Before all this was arranged, I had organized a dinner party at my home. Gene and I had invited my old boss, Doc Draper, along with Don Hornig, Johnson's science advisor, and a few other people. Jim Webb had encouraged me to stay home and host Doc Draper, who was playing an important role in the Apollo program.

I arrived home a little bit before seven o'clock in the evening. As I opened the door, the phone was ringing. Moments later Gene called and said, "Is that you, Bobby? It's George Low on the phone." I picked up. At first I couldn't understand what George was saying. It was something like: "They're all dead."

"Who's dead?," I asked. Then he named the three astronauts. I got some further particulars from George, then said to him, "I'm going to my office. I'll get back in touch with you from there to find out a lot more about what's going on."

I told Gene that we had had a serious accident. I said that I thought she ought to have our guests come to the house, have the dinner party, and not tell them what was going on. I felt that if, as they came in the door, she announced what had happened, it wouldn't be much of a dinner party, and our guests wouldn't know whether to stay or to go home.

When I got to the office, I went about the business of trying to be sure I knew the facts and making sure that everybody who should know about it did know. I started getting calls. One was from Defense Secretary Robert McNamara's office. He had heard something about the accident and wanted to know the particulars. While I was talking with him, an operator cut in, apologized, and said she had an emergency message. The "emergency" turned out to be Peter Hackes of NBC News. He said, "This is a national emergency. The word is out. The country is almost in a panic, and you've got to go on TV and reassure the public!"

"That's ridiculous," I said. "How can I do that? I don't know all the facts. No, I'm not going to appear on TV." I probably would have not said the same thing if we hadn't been through the crisis on Gemini 8, but that incident had taught me that in an emergency, public information had to be released in a considered, careful way. This event was of such a magnitude that I knew only Jim Webb had the authority to release such information.

Nothing could be done until we knew more about the accident. We had to grab hold immediately of all of the pertinent data and equipment for analysis. I hauled out the disaster plan that I had reviewed only a few months before and saw that one thing had to be changed: the accident review board would have to report directly to Mr. Webb, not to me, as the plan indicated.

By midnight George Mueller and I had put together a list of people we felt should be on the review board. Clearly it had to include an astronaut. Not only were the astronauts test pilots with technical degrees and broad experience within the program, but also they would want assurances from one of their own before proceeding with space missions. Frank Borman proved to be an excellent choice. We wanted to have a lawyer on the board and chose George Malley, the chief counsel at Langley. On down the list we went. But who should be the chairman? Clearly he, like the others on the board, should not have had an active role in the running of the project. Yet we also wanted to keep the review within NASA as much as possible. Floyd ("Tommy") Thompson, who ran Langley, was a person who would have everybody's respect and confidence. We chose Tommy.

I arranged to have one of the NASA planes pick me up in Washington the following morning at 6:00 a.m., then fly on to Langley Research Center for Tommy Thompson and George Malley before heading to Cape Canaveral. There I made sure everybody fully understood the ground rules—that the review board had complete authority over the investigation, that they were to select the consultants and to determine who at NASA worked on the review, that they were responsible for all of the documentation, and so on.

At the Cape, I met with Sam Phillips, Kurt Debus, Joe Shea, and three or four others. They told me what they knew about the tragedy. We talked about the dead astronauts' families. It was decided that I would be the one to take a look at the pictures of the capsule taken just after the door had been opened, since someone in authority needed to do so. When, on my next visit to the Cape, I met with the medical examiner and reviewed the photographs, the experience wasn't pleasant, but it wasn't as gruesome as might be imagined either. Ash from the fire had settled over everything in the capsule, including the astronauts' bodies, and the three of them were roughly in the same positions they had assumed before the fire broke out.

In the Capital, plans had been made for the various funeral services. I accompanied the First Lady and the Vice President to West Point, where Ed White was buried. Jim Webb, together with his wife and mine, accompanied the President to Arlington, where Gus Grissom and Roger Chaffee were laid to rest. By this time, Jim had worked out an agreement with the President that NASA would handle the review and that there would not be a presidential commission. We recognized that information on the review needed to be transmitted to the President, to the Congress, and to the public as the review proceeded, without limiting the options of the review board. Jim's solution was sound. It was decided that the board would not issue any kind of report before its final report. I was given the job of going to the Cape every week and meeting with the board. They gave me the facts as they knew them at the time; then I reported in writing to Jim Webb as soon as I got back to Washington. He personally took my report over to the President, and within a few hours it went up to the key chairmen on the Hill. The document was then released publicly. Importantly, it was never signed by Tommy Thompson, so that his and the board's hands would not be tied when it came time to file their formal report.

By the time I submitted my first report on the progress of the review board, Jim's demeanor clearly was not the demeanor of the Jim Webb I had been working with for the past six years. He was right that my report had an emotional tinge to it, with some expression of concern for the families. But the way he slashed those sections out! I was stunned. Even when I rewrote the report as a strict chronology of events and presented it to him, he didn't say, "Nice job." He was gruff. I hadn't seen him that way since the last days of Brainerd Holmes, when Jim knew he had a difficult personnel problem on his hands.

Everyone at NASA was feeling the strain. The trouble was not that our people didn't care enough about the fire; they cared too much. Key people from Houston would fly up to Washington to testify and literally sob all the way on the plane. I first realized the seriousness of this problem when I was down at the Cape for the two-day review of the preliminary findings and recommendations. Certain individuals seemed in very great need of sleep. I didn't fully appreciate the extent of the problem until the next day when I was flying back to Washington in the company of Chuck Berry, NASA's chief medical officer. He explained to me that he had been ministering not only to the astronauts and their families (particularly those of Grissom, White, and Chaffee), but also to quite a few people in the NASA organization.

Joe Shea was the most affected. As the person in charge of the Apollo capsule, he was bound to take personally the fact that this had happened on his watch. In addition, if not for a fluke, he would have been in the capsule with the three astronauts, observing their work, when the fire broke out. He had planned to lie at their feet during the test so that he could follow their communications. At the last minute, his headset didn't work. He said there was no point in being inside with them if he couldn't hear what was going on, so he got out of the capsule and flew back to Houston, where he received the grim word.

I knew Joe well. We played a lot of tennis together. I could see that he was extremely upset. Jim Webb went way out of his way to be helpful to him. After leaving NASA, Joe went on to become vice president for engineering at Raytheon and served on several important committees of the National Academies of Science and Engineering. In 1990, he became an adjunct professor at MIT, where ultimately he and I shared an office.

Back to the Hill

Needless to say, Congress wanted to be involved in investigating the accident. We had several sessions on the Hill. For the first, an executive session of the Senate, it was agreed that I would appear with George Mueller, Chuck Berry, and others, and that right afterwards I would step outside with Senators Clinton Anderson (D-NM) and Margaret Chase Smith (R-ME), the committee chairman and ranking minority member, to summarize for the waiting media what had taken place. I knew when I walked into the hearing room that this press appearance would be lively because I had never before seen more coaxial cable strewn around a corridor.

Just before the end of the session, Senator Anderson left the chamber. Of course, he was caught by the networks, but I didn't know what he had said. When we were through, I went to Senator Smith and asked her if she wanted to join me in meeting the press. She was pleasant but made it clear that she thought it really would not be appropriate for her to do so. I walked out of the hearing room alone into blindingly bright lights. Not knowing what Anderson had said, I had to be careful not to run the risk of contradicting him. The press, as always, was looking for a story and would have loved to have caught me in a misstatement. I don't think any great harm was done by what transpired there, but it was one of the livelier moments in my NASA career.

The next session in front of the Senate was thrown open to the media. The newspaper people were all behind us at the witness table, and the TV people were free to turn on their equipment in front of us at any time. Each of the senators on the panel questioned us. Senator Walter Mondale finally had his turn. He asked whether, prior to the accident, there had been a report critical of North American Aviation and possibly recommending that we change contractors.

George Mueller said no, there had not been a report. I knew George meant that there hadn't been a formal, bound report. But I suspected that Mondale may have had in his possession a version of the results of an informal "tiger team study," led by General Sam Phillips. The tiger team had never issued a report, per se; however, the team had prepared briefing charts in order to discuss perceived problems with North American. So in order to leave a crack in the door for future discussion (I was thinking of *immediate* discussion after the session with Mondale and the chairman of the committee), I interjected a few thoughts about the kind of review we normally carry out. I added that I wasn't aware of any specific report, which I wasn't. As far as I knew, there had never been a recommendation to go to another contractor for the spacecraft.

Jim Webb nabbed me outside the committee room and asked me to drive back with him and our general counsel, Paul Dembling. When we got into the backseat of his car, the first thing Jim did was crank up the window separating us from the driver. Then he lacerated me. "You don't volunteer information!," he said. "You can't look at these as ordinary hearings, like any we've had in the past. You've got to look at these as legal proceedings. Don't volunteer information unless you're sure. Don't volunteer information, period!"

"Well," I said, "I just thought there might be some kind of report."

"Don't speculate in front of Congress!" Before we got back to headquarters, I was able to say that I was very concerned about the Phillips study, that I felt it was the kind of thing that could cause very great trouble, and that I thought the way to solve it was to get the facts up to Senator Anderson and others about what had really been done by the Phillips task force.

Back in the office, I was doing a slow boil when Paul Dembling came in and showed me the Phillips document. Copies of the briefing charts were bound together, and a letter signed by Sam Phillips was appended. "This report," the letter read, in effect, "was made after a careful review of management practices at North American."

I scowled, "Take it in to the boss." It was a bad time, and it would get worse.

Jim was forced to acknowledge that, contrary to our testimony, there had been a report. The next Senate session fell on the same day as our appropriations hearing in the House. Jim took the House session, and I took the Senate hearing, which again was open to the media. Representing NASA were General Phillips, George Mueller, Chuck Berry, and myself. I had a good chance to bring up the Phillips study and to have General Phillips discuss it in detail. He did an excellent job of presenting the material, cold turkey.

Senator Mondale asked of the report, "Well, shouldn't all of this have been released?"

I answered, "If every time we had an internal review, everything was released, good internal reviews would become unachievable. If they are made with the idea that they will be discussed on national TV, they will never contain controversial information. I agree, however, that the results should be available, and we have discussed them here in this hearing in a very open way." Indeed, we had.

At this point there was a roll call, and the senators all left the committee room to vote. As they came back in, Mondale walked right by me and said, "Well, I know some of your problems in the spacecraft." He was obviously referring to the crowded room with cables all over the place. I was not exactly filled with mirth, but I did smile at his comment. The next day the *Washington Star* carried a picture of Dr. Seamans testifying at the Senate hearing on the tragic Apollo fire—and smiling.

Watershed

The Apollo 204 fire was tragic, and I have often asked myself whether it could have been avoided. The use of a 100-percent oxygen atmosphere at sea-level pressure accelerated the burning so that it could not be extinguished once ignited. A single-gas system had been used in both Mercury and Gemini capsules, and for rational reasons. Extensive testing had been carried out to find capsule fabrics that were most fire-resistant; however, while fire retardants and extinguishers had been examined, none had been found adequate to put out a fire once started.

What more might have been done? Comprehensive rocket testing was conducted by engine contractors and by the government at the Mississippi test facility. All manner of hardware and system tests were run under a wide variety of conditions. But a boilerplate capsule was never used to investigate the incendiary nature of a pure-oxygen fire at sea-level pressure. If such a test had been conducted, it is almost a certainty that the design would have been changed prior to the fire of January 27, 1967. Neither I nor anyone else, to my knowledge, suggested such a test.

This oversight led to tragedy, but the fire could have been much, much worse. We could have killed everybody on the pad outside of the capsule. On top of the capsule was the so-called escape tower with rockets in it, designed to take the astronauts away in the capsule if something happened just before or after liftoff. If these rockets had been triggered (the fire did scorch the outside of the capsule), the whole pad might have been destroyed. I'm not certain the Apollo program could have recovered from such a disaster. I have also thought that if we had not had the Apollo 204 fire when we did, we might well have failed in our overall mission of going to the Moon by the end of the decade, because a similar catastrophe might have hit us later on, when it would have been much more difficult to recover.

In certain respects, the impact of the fire was similar to that of the assassination of President Kennedy. There was tremendous shock in both cases. In both cases it was hard for people to accept the fact that it had happened. In both cases there was strong public identification with the individuals. Of course, this was much greater in the case of President Kennedy, but Grissom was well known to the U.S. public as one of the first seven Mercury astronauts, and Ed White was quite well-known and greatly admired for his "spacewalk" during Gemini 4. Chaffee, who had never flown a mission, was relatively unknown. The public had also built up NASA as infallible. We had done miraculous things. Now suddenly we had made what seemed an inconceivable mistake. It would have been one thing if the fire had occurred in space, but on the pad? And not being able to get the men out! No one could understand such a colossal blunder.

Understandably, Jim Webb took all this personally. He became terribly tense. Migraine headaches, which he tended to have anyway, were exacerbated. I had the feeling I was dealing with somebody who could explode at any moment. I'm not a psychiatrist, but I would say that the fire came as a tremendous psychological blow to him. Before the accident, he and his program were riding high. He was front and center, getting acclaim from many, many quarters and deserving it. Given his age, sixty, Apollo would probably be the last major endeavor of his working life. It would be a fitting monument to his ability.

Now his house of cards was down. How? Why? Who had made the mistake? Who had destroyed his dream? It was necessary, of course, to carry out a complete and careful investigation, so that the engineering failures that had led to the fire could be corrected. But Jim was not interested in investigating the engineering. He wanted to know what individuals had failed him. He felt personally betrayed.

There was no question in his mind that North American, the contractor for the Apollo capsule, was one of the culprits. In particular, Jim felt that North American's project leader, Harrison ("Stormy") Storms, had failed him. When we selected North American, Dutch Kindelberger, a hard-hitting, forceful manager, had been the chairman. By the time of the accident, Lee Atwood had succeeded him. We were soon having meetings with Atwood. They were not pleasant meetings. Jim Webb made it clear that changes were going to have to be made, to which Atwood responded, "Let's not panic! We've had accidents before. We're not part of the government. We're a separate institution. We have to manage things the way we believe is right." In effect, he was saying, "You're not going to dictate terms to us." Partly at my suggestion, we had conversations with other potential contractors, so that North American would realize that although they had the contract, we wouldn't necessarily continue with them. If they wanted to dig in their heels, we would dig ours in, too. Finally North American did agree to make substantive changes. They took Harrison Storms off the job and agreed to a $10 million reduction in their fee. Also, Boeing was brought in to be the systems integrator. That was an important move and one that we should have made earlier, regardless of the fire.

North American aside, there was no question in Webb's mind that people inside NASA also had failed him. He clearly had lost confidence in the ability of the organization. He started talking about George Mueller's shortcomings. I told Jim that George had deficiencies like everyone else, but at the same time, George had made many positive contributions. Pretty soon I started to hear that Jim was talking about my imperfections behind my back. As time went on, I found that assignments were being given that I didn't know about. At a meeting on the Voyager project,[4] which was still being formulated, a report was presented by Mac Adams, the associate administrator for research and technology, about which I had been told absolutely nothing. As general manager, I had been Jim's line of communication to the organization. Now he was bypassing me.

"Jim," I said at the Voyager meeting, "if you don't mind my saying so, I think we'd make a little more progress if you'd let me in on some of these studies."

He froze, then turned on me and said, "No more of that kind of talk, Seamans!"

Jim Webb's reasoning was a little like a geometric theorem. He was a nontechnical person and believed that the technical staff had let him down. As de facto general manager, I was his bridge to the technical people. Therefore, the bridge had failed and needed circumvention.

In one notable instance I did fail badly: I had had discussions with the press that I shouldn't have had. By this time I had considerable experience dealing with the media. I knew the media weren't perfect, but I also recognized that there were many good journalists. I also

---

[4] The Voyager Mars mission then under discussion was canceled in the fall of 1967 as a result of congressional budget cuts. Several years later the project was revived as the Viking program and landed two spacecraft on Mars. The name "Voyager" was reused for the missions to Jupiter, Saturn, Uranus, and Neptune, beginning in 1978.

thought that Jim Webb was too paranoid about the media and that it would be helpful if a few responsible journalists were more knowledgeable about what was going on in the accident investigation. The media were hungry for information about the fire. All they had was what they could get out of the congressional hearings and my reports after visits to Cape Canaveral.

Julian Scheer, NASA's public affairs officer, came to me one day while Jim was out of town and said, "I'm thinking of inviting" (he named eight correspondents) "to come in for a little background on what's really going on." A *backgrounder* in press parlance is an interview not to be quoted but meant to give the media some insight. I felt that what Julian was suggesting could be quite positive. So we invited the correspondents to a luncheon. If Webb had been in town, probably we would not have done so.

It was an informal lunch, and the reporters asked questions like, "Why couldn't you get the astronauts out of there?"

I explained that we had designed the door to open inwards to avoid a mistake in space—an astronaut putting his elbow in the wrong place and suddenly losing all cabin oxygen. With the pressure that the fire created within the capsule, it would have taken an 8,000-pound pull to get the door open. The next day papers ran stories about the astronauts struggling to get out, clawing at the door—all of it horrible stuff greatly exaggerated. In fact, evidence showed that the astronauts had not burned to death, as most had originally assumed. They had suffocated and were unconscious no more than seventeen seconds after the first spark.

When Jim Webb returned to the office, he was beside himself. Julian and I had let him down. He had had a handshake with the President and the Congress that no information would come out without their getting something in writing first. From this point of view, I was clearly in the wrong. From then on my relationship with Jim Webb went almost straight downhill. It became obvious as the summer of 1967 evolved that our lack of rapport was not a good thing for the organization. I had been considering the possibility of leaving NASA prior to Hugh Dryden's death, because I had been there already twice as long as I had planned initially. I had also been asked to consider the presidency of a well-known university. I finally had decided, a few months before Hugh died, that it would be inappropriate for me to leave at that time.

Now it was clearly time to leave. I felt the need to be with my family more, to begin a new professional life, to have a chance to relax and regain my perspective. The program that I had come in to work on, Mercury, had long since ended. I felt (and would always feel) very much a part of Apollo, but I also felt that to stick around for the sole reason of being there when we went to the Moon was the wrong way to make one's personal decisions.

I wanted to get out in such a way as to do the most good to NASA and the least harm to myself. This proved to be quite easy. I got in touch with a good friend of mine, Walter Sohier, who had been general counsel to NASA, and with my brother Peter, who is a lawyer. Together the three of us figured out the best possible exit. We sat in the warm autumn sun on the third-floor deck at Dumbarton Rock Court and composed a letter of resignation. Gene typed it. The next day, October 2, 1967, I handed it to Jim Webb.

Jim looked at the letter, then at me, and said, "What do you think your peers are going to say about the job you've done over the years here at NASA?"

"I think they'll feel that I did a satisfactory job."

He got up and left the room. He went immediately over to the White House to see President Johnson, and within just a few hours it was announced that I was leaving NASA.

Afterthoughts

After the announcement of my resignation from NASA, Jim asked me to stay on for three months, full time, which was fairly unusual. At no time during this period did I take public issue with him. I never said anything substantive to the press about my real reasons for leaving. I was invited over to the Washington Post to meet with their senior editors, who asked, "There must be some reason why you're leaving right now. Why not stay until you get to the Moon?"

I said, "Look, I've been down here seven years. I only intended to stay two. It's been pretty hectic, but we've got everything pretty well in place. What's the point in sticking around for some kind of big group ceremony?" They accepted that explanation. If I had taken on Jim Webb openly, if I had left NASA making reckless statements, it would have hurt NASA, and I'm quite sure I would never have been asked to return to government service. By leaving without a confrontation, I left the door open for the future.

Jim made quite a big deal about swearing me in as a consultant on the day I officially retired. By that time, his demeanor toward me had started to change for the better. Not long after I left, I was given the Goddard award for my contribution to the space effort. President Johnson presented the award to me at the White House with Jim in attendance. Then Jim indicated that he wanted to have a little going-away party for me and asked me whom I wanted to invite. The date of the dinner happened to fall immediately after the assassination of the Rev. Dr. Martin Luther King, Jr. Protests gave way to riots, and Washington, D.C., became an armed camp. Since a dusk-to-dawn curfew was in effect, we had the dinner at the Army-Navy Club and held it early enough so that we could all return home before the curfew.

Dr. Thomas O. Paine, a materials engineer from General Electric, replaced me as deputy administrator. With the presidential election coming up in the fall of 1968, Jim Webb, who was tired after seven and a half years on the job, decided it would be a good time to retire. He thought that by resigning before the election, he would give President Johnson an opportunity to install Tom Paine as administrator and that the new administration might keep Paine on. This, Jim thought, would help maintain some continuity in the NASA effort. On September 16, 1968, Jim went in to see President Johnson and said he had been thinking about resigning early. "I've been thinking along the same lines, Jim," the President said. "Let's step outside and tell the press that you're leaving, effective immediately." Suddenly, Jim was gone. Johnson was always direct and to the point!

In March 1969, soon after I had become secretary of the Air Force, I got a call from Defense Secretary Melvin Laird. "President Nixon," he said, "wants to know my views on keeping Tom Paine at NASA, not as acting administrator but as the administrator. What do you think?"

I answered, "Well, I can give you a very straightforward, simple answer, Mel. Ask the President if he wants to carry out the lunar landing this year. If he does, make Tom Paine the administrator. But if he wants to run the risk of not going this year, then bring in somebody else." The next day the President announced that Tom Paine was his nominee for administrator. Jim Webb's strategy had worked, and the first Moon landing took place four months later.

My view of Jim Webb and of our last year together at NASA changed over time. I still think I did the right thing to leave when I did, for the good of all parties. But it was only after I had left and time had elapsed that I could fully appreciate the superb job Jim had done over the years. He made it possible for NASA to do what surely is one of the most difficult technical jobs ever accomplished. It's hard to dream up a model manager who could have come in and done as well as he did. Later, as secretary of the Air Force, I came to realize that my job wasn't comparable to Jim's role at NASA, because I always had the secretary of defense available when trouble developed. But when I got involved with the Energy Research and Development Agency (ERDA) in 1974, I saw that I had no place to turn but the President of the United States or the Congress, and I got a much better understanding then of the challenges Jim Webb had faced.

Fortunately, I had a chance to express my views at a dinner party given by Dave and Pat Acheson. Dave was the son of Dean Acheson, Truman's secretary of state, for whom Jim had worked. Gene and I were invited along with Dave's mother, Jim and Patsy Webb, and six to eight others. It suddenly occurred to me sitting there at the table that it might be appropriate to make a toast to Jim Webb and the job he had done at NASA. I said quite sincerely that I had never truly appreciated all that Jim had done until I became administrator of ERDA.

About a year later, Gene and I went to see the Webbs. Afterwards, Gene told me that Patsy Webb had taken her aside and had said, "You'll never know how much that meant to Jim, when Bob said what he did about the role Jim had played at NASA. You may not know it, but Jim didn't always have an easy time with Mr. Acheson in the State Department, so for Bob to have said what he said in the Achesons' house was doubly gratifying."

In the mid-1970s, Jim Webb was diagnosed with Parkinson's disease, and I periodically went around to see him. We would sit and chat, and pretty soon it was like the old days. Over time he got progressively worse. Sometimes he would be wheeled into the room or drive himself in in a little electric cart. At other times he walked, but with great difficulty. After a while he couldn't focus his eyes and wore a patch over one of them. Invariably, though, he had a report, article, or book to show me. "I think you'd find this very interesting," he would say. "I'd like to know your thoughts on it. Drop me a note and tell me what you think"—as though I were still working for him!

The next-to-last time I saw Jim was in 1991. He had just had his eighty-fifth birthday and was hospitalized. When I came into his room, his wife, Patsy, immediately left so the two of us could chat. He had a lot of things on his mind about problems with the government and the country and things that I might do to help out. He was trying so hard to express himself that his whole body was almost writhing with the effort. As I got up to leave, he said, "Bob, I'm not going to live much longer, but I'd like to know what you hope to accomplish before you kick the bucket."

I wasn't quite sure how to respond to that, except to say, "Hey, Jim, you may be around for a long time."

But that wasn't in the cards. I saw him once more at his home. He looked better than he had in the hospital, and he was still bubbling with ideas. Not long after that he had a fatal heart attack.

Mission Accomplished

Gene, Joe, and I flew down to Cape Kennedy for the launch of Apollo 11 on July 16, 1969. The night before the launch Gene and I were invited to a small dinner hosted by the former President and Mrs. Johnson. I made a few remarks about how much the program owed to Johnson's support.

After the launch, I went around to launch control to congratulate everyone for getting Armstrong and company off to a good start. I found Vice President Spiro Agnew there, making a speech about the future of NASA. Being far more political than technical, Agnew must have seen that there was a lot of glory in space and figured he wanted to be a part of it. He told the crew on hand that this was just the start. Before we knew it, he said, America would be going to Mars. I was very skeptical, to put it mildly.

We flew back to Washington later that day; then about three days later, we flew down to Mission Control in Houston for the Moon landing. We sat in a glass-enclosed observation room behind the control consoles. The landing was a close call. The computer was periodically overloaded because a radar switch had been inadvertently left on, but Mission Control decided to override the error signal (wisely, it turned out). Then the lunar lander came within a few seconds of running out of fuel for its descent to the surface. After landing safely, Armstrong and Aldrin had a rest period of six or seven hours before stepping out onto the Moon. A bunch of us went out to get something to eat nearby while waiting. With us at dinner were Doc Draper and Jackie Cochran, the world-famous aviatrix.[5] We finally went back to Mission Control and watched the astronauts walk on the Moon. That was incredible. I could hardly believe it.

I did not feel any regret about not being a part of NASA at this historic moment. By this time my plate at the Air Force was very full. I couldn't help but feel excited though, and I certainly felt proud not so much of my own accomplishments but of NASA itself and of the many gifted people who had made the Moon landing possible. People like the astronauts. People like Sam Phillips and George Mueller. People like Rocco Petrone, who had run the massive construction projects at the Cape. I felt great pride in taking my son Joe and other guests around the Cape and Mission Control and explaining how the whole fantastic system worked.

I can't help but look at the Moon today and think it's amazing that we were there more than twenty-five years ago. I believe we will return there someday, but mounting that kind of an effort is not something that's going to happen, in my view, for a long time, because there's not enough reason for doing it. As for Spiro Agnew's prediction that America would fly to Mars, I learned a long time ago that if you say something will never happen, you'll eventually be proven wrong.

---

[5] Later in my tenure as secretary of the Air Force, I was on hand to honor her when she retired from the service. Afterwards, Gene asked her if she had competed in the Powder Puff Derby, a cross-country race for women pilots. Jackie Cochran drew herself up with horror, sniffed, and said, "I only compete with men!"

Grandpa Bosson pointing out my first airplane, 1922.

Some thought we were too solemn during the ceremony,
but not outside St. John's Parish, June 13, 1942.

On November 11, 1962, President Kennedy and Vice President Lyndon Johnson visited the Marshall Space Flight Center. Werner von Braun, the center director, points out a large model of the Apollo–Saturn V, with an H-1 engine in the foreground. From right to left: Kennedy, von Braun, Johnson, myself, and Paul Nitze, the assistant secretary of defense for international security affairs. The two men at the far left are unidentified. James Webb and Robert McNamara are obscured behind Kennedy and von Braun. Edward Welsh is wearing dark glasses, in the center background.

I take a ride in the Gemini flight simulator with astronaut Neil A. Armstrong. In August 1963, I visited the Manned Spacecraft Center in Houston for a review of operations. NASA photo number S-64-31305

President Kennedy visited Cape Canaveral on November 16, 1963. George Mueller briefed President Kennedy in the Complex 37 blockhouse. From left to right: George Low, Kurt Debus, myself, James Webb, Kennedy, Hugh Dryden, Wernher von Braun, Maj. Gen. Leighton Davis, and Senator George Smathers from Florida. Models of the Saturn V rocket on its crawler/transporter and of the Vehicle Assembly Building are on the table. Several periscopes hang from the ceiling.
NASA photo number LOC 63P-136

President Kennedy inspected the Complex 37 launch facilities with Wernher von Braun (center) and myself. Kennedy points at the live second stage of a Saturn I rocket. Von Braun used the model in the foreground to explain the powerful rocket's mission. Six days after this photograph was taken, President Kennedy was assassinated.
NASA photo number LOC 63P-145

The NASA management triad of Administrator James E. Webb (center), Deputy Administrator Dr. Hugh L. Dryden (left), and myself as Associate Administrator. This photo was taken in Webb's office. NASA photo number 66-H-93

James E. Webb swears in NASA's new deputy administrator, February 4, 1966. At left is Vice President Humphrey.

The November 22, 1966, postflight press conference on the Gemini XII mission. From left to right are myself, astronaut James E. Lovell, astronaut Edwin E. "Buzz" Aldrin, and Dr. Robert R. Gilruth, the director of the Manned Spacecraft Center. NASA photo number S-66-65190

On October 24, 1967, I had the privilege of presenting a NASA award to my early mentor, Dr. Charles Stark Draper.

In the Oval Office, fall 1969. From left to right: National Security Advisor Henry Kissinger, Secretary of the Army Stan Resor, Secretary of Defense Mel Laird, President Nixon, Secretary of the Navy John Chaffee, and myself.

*Four F-15 airplanes
flying in formation.*

*The A-10 aircraft in
flight over a prairie.*

An F-111 at Langley Air Force Base, Virginia.

After an F-111 flight over the Nellis Test Range in Nevada. We flew at Mach 2 at high altitude, then used Automatic Terrain Clearance at subsonic speed to fly over hills at an altitude of only 300 feet.

Showing President Ford the seal of the new Energy Research and Development Administration (ERDA).

I watch as James C. Fletcher, NASA Administrator, signs a cooperative agreement between NASA and ERDA on energy research and development.
NASA photo number 75-H-678

As head of ERDA, I took a trip with Admiral Hyman G. Rickover many feet below the ocean surface in the nuclear attack submarine Cravallus.

With Vice President Dan Quayle at the June 1991 release of a study on NASA future missions by the Space Exploration Initiative team. I served as vice chairman of this study panel.

CHAPTER THREE

The Air Force Years

By March 1968, I had a joint appointment at the Massachusetts Institute of Technology (MIT) in the Sloan School of Management and the engineering school. The family continued living in Washington, and I commuted to Cambridge to give seminars and so on. In May we bought a house in Cambridge, and in the summer we came north to live in our house at Beverly Farms while the new house was renovated.

By fall I had a more formal appointment at MIT, as Hunsaker Visiting Professor for the academic year 1968–1969. The only requirement of the Hunsaker professorship is that one deliver what is known as the Minta Martin lecture. The recipient is also invited to participate on thesis committees, to give seminars, and so on, which I did and enjoyed. We lived in the third-floor apartment of the house while massive reconstruction went on in the basement and on the first and second floors. As a result of all the old plaster in the air, an asthmatic condition that had bothered Gene in Washington intensified. Over Christmas she had to be hospitalized for a short time. She wasn't allowed to go back into the house until all the work was completed, so we moved temporarily into an apartment hotel. After the extensive renovations, our house was comfortable, sunny, and spacious without being too big. Gene even had a little greenhouse. The house's location was very convenient, and it answered our needs perfectly. We were a bit sad when we finally gave the key as a donation to MIT on April 1, 1992. The proceeds from the institute's sale of the house provided nearly 50 percent of the funds required to fund MIT's newly created Apollo Chair in Astronautics.

We found a buyer for our Washington house in December 1968,

just when we realized we might be moving back. I was at MIT the week before Christmas, when my secretary told me that newly appointed Secretary of Defense Melvin Laird was on the line. He introduced himself and asked if I would be in Washington soon. I was planning to be there the following day, en route to Cape Kennedy, where I was going to watch the launching of Apollo 8, the first circumlunar flight.

Laird said, "I'd like very much to have lunch with you at the Carlton." We had lunch in his private suite at a table for two. One whole wall was covered with a detailed organization chart of the Pentagon. As we began eating, he asked me questions about technical people. I figured he only wanted information, but by the time we got to the raspberry sherbet, he shifted and said, "Dave Packard [co-founder of Hewlett-Packard] is going to come and join with me as my deputy. I want to have three service secretaries with different backgrounds. I'm hoping Stan Resor, secretary of the Army, will stay on from the Johnson administration. Goldwater thinks a lot of John Chafee from Rhode Island [who had recently lost his reelection bid for governor]. I want to get him down here for secretary of the Navy." Then he said, "And you're going to come down here as secretary of the Air Force."

"You've got to be kidding! I just got out of the government and moved back to New England. I've just begun a new job at MIT, and my wife is sick." Laird absolutely would not take no for an answer. In fact, he upped his offer, saying that Dave Packard could not stay more than two years and that after Dave left, he wanted me to step in as deputy because, like Dave, I had a technical background.

"Why don't we get together again on Monday?," he said.

I didn't have the courage to call Gene until I had reached the Cape. When I finally did call, her initial reaction was: "We can't do that!"

"Don't worry," I assured her, "I'm not going to accept."

I saw Laird again on Monday, and we had a further conversation, but I still didn't agree to take the job. "Mel," I said, "if you don't mind, I'd really like to take until the day after Christmas. I'll call you on the twenty-sixth of December."

I went to see MIT president Howard Johnson during Christmas week. "Well," he said, "when the President knocks on your door, it's pretty hard to turn him away."

"They'd like me to start on the twentieth of January. How do you feel about the Minta Martin lecture?"

"It's the custom to deliver that lecture." I was still on the hook, but fortunately I already had done some work on the lecture.

Waiting until after Christmas to give Laird an answer gave me a chance to talk to our four older children separately. I asked them what they thought I ought to do. They said they didn't think much of the Department of Defense because of the ongoing Vietnam War but that if anyone had to help run it, they would just as soon it were me. On December 26, after a final consultation with Gene, I called Laird and accepted.[1]

We passed papers on our Washington house on the very same day. We could have stopped the sale, paying some kind of penalty, but by then the thought of moving back into that house didn't have much appeal. It would have meant moving back all the furniture that we had recently lugged north to Cambridge. Also, by then parts of Georgetown had changed, with mobs of people and drugs galore. Instead, Gene suggested renting a furnished house, which we did. We stayed in a rented house for two years. When it became apparent that we were going to stay in Washington longer, we bought another house and moved our furniture from Cambridge back down to Washington.

On January 8, 1969, the three new service secretaries—Stanley Resor (Army), John Chafee (Navy), and Seamans (Air Force)—were trotted out for public view. I wasn't sworn in until February 15, having received an extension from Laird in order to finish the Minta Martin lecture for MIT—though I did meet with Harold Brown, outgoing secretary of the Air Force, several times before then. The swearing-in occurred on a Saturday morning at the Pentagon.

---

[1] I never did become Laird's deputy. Mel named me for the post after Packard's two-year hitch, then nominated Curtis Tarr, head of the Selective Service System, to move into my office at the Air Force. Word came back from the White House that they didn't like this change. It was finally decided to make Ken Rush, our ambassador to Germany, Mel's deputy, and I remained secretary of the Air Force.

Forming the Team

By February 15th I was well along in my selections for the Air Force secretariat, but none of the key jobs had yet been filled. The presidential appointees reporting to the secretary were the under secretary and four assistant secretaries (Research and Development, Installations and Logistics, Financial Management, and Manpower and Reserves). I was most grateful to Harold Brown for staying in his position until February 15th and assisting in the transition. The morning I was sworn in, Harold departed to take over the presidency of the California Institute of Technology.

On one of my Washington visits in January, I met Glen P. Lipsomb who was a Republican congressman from California and a great friend of Mel Laird. He had a long list of individuals who had been suggested for presidential consideration. The list had been screened by him and the White House. I recognized some real dogs and wanted to be certain I didn't have to take time to interview them. But one person on the list was Grant Hansen. Grant had served as project director of the Centaur launch vehicle, the first hydrogen-fueled second stage, designed to mate with the Atlas booster. When he took over, the project was in disarray and NASA was about to cancel General Dynamics' contract. He saved the project, and the Centaur is still in use today. I found he was about to leave California to take an advanced management course at Harvard. I convinced him to come to Washington instead.

Harold Brown had a generalist named Tim Hoopes as his under secretary, which placed a heavy load on Al Flax, his research and development (R&D) assistant secretary. Al was responsible for all R&D, including the C-5 and the F-111 aircraft, and in addition was responsible for the highly classified National Reconnaissance Organization (NRO), which operated the nation's observation satellites jointly with the CIA. I elected to transfer the NRO to the under secretary and hence was looking for a second technically oriented individual. Harold Brown had given me a list of fifteen names he felt were qualified for the R&D function. One of them was Dr. John McLucas, president of MITRE, a nonprofit corporation located near Hanscom Air Force Base outside of Boston.

Back in Massachusetts, I went to see James R. Killian, Jr., chairman of MITRE's board and former chairman and president of MIT. I said I wanted to talk to him about Dr. McLucas since there was a possibility of his coming to Washington. As I watched the color drain from his face, I realized that I was considering an excellent candidate, which I really knew anyway. But he properly questioned whether, with all of Dr. McLucas's experience, it would be appropriate for him to come down to be the assistant secretary. So I said, "Well, what would you think of him as the under secretary?" He said, "I would think that would be a very appropriate job for him, but understand, I hope he doesn't accept. But you'll have to speak to him about it." So on one of my trips via Hanscom Air Force Base, I asked for an office, and Dr. McLucas and I met. I made my proposition that he become the under secretary of the Air Force and head up the classified programs. He accepted almost on the spot but said he wanted to talk to his family prior to making an absolute commitment.

Grant Hansen and John McLucas provided great strength in research and development. The secretariat also needed capability in administration, finance, and personnel. Although Philip Whittaker and I had not overlapped at NASA, I knew he was highly regarded both there and at IBM where he had worked previously. In addition, he had been most helpful obtaining information for my Minta Martin lecture at MIT. So I cast my net and fortunately landed him.

Spencer J. Schedler did have some political experience. He was an advance man, as it turns out, for Spiro Agnew. That's how the name came to me, but it had nothing to do with why I picked him. I selected him because he liked the Air Force and was still flying in the Air National Guard. Some people were amazed when I made clear that "I want somebody who knows something about finances, how to keep books and how to run audits." He was a graduate of the Harvard Business School and came highly recommended. He was the youngest of the group.

Finally, there was Curtis W. Tarr who was the assistant secretary of the Air Force for Manpower and Reserve Affairs. Tarr had been president of Lawrence University located in Appleton, Wisconsin, and had also worked as a junior member of the Hoover Commission on the reorganization of the Department of Defense. His doctoral thesis at Stanford was written on the subject. I was really looking for somebody who had both organizational as well as a personnel background. Now he wasn't really a personnel man as such, but he had certainly grappled with organizations large and small and on personnel problems during a troublesome period for universities.

I found in my first six months that we were taking a shellacking in the political environment of the Department of Defense. We didn't have to go up to the Hill to run into politics. There are vested interests in the OSD (Office of the Secretary of Defense) staff, and in the military services as well. We weren't always coming out very well in this tug of war.

So there was something missing in our secretariat. Jack L. Stempler had been a Marine in World War II, a civilian lawyer in OSD, and most recently the head of OSD's legislative affairs office. Mel Laird suggested that he become our general counsel. He had probably served his usefulness as legislative liaison at the OSD level—that's a job that no human being can possibly take on for many years because he's using up his chits, all the time. We had a good general counsel named John M. Steadman, whom I liked and who was the only holdover that we had incidentally. But Jack Stempler obviously had a background that would be extremely helpful because of his political savvy on the Hill and, really more important, his understanding of the Department of Defense. He was the missing link, if you will, that I had been looking for in rounding out the secretariat. He was what Dr. Brown would have called the "generalist." I had learned from NASA that if you get the right kind of a lawyer to work with you, not on legal matters—though obviously he has got to get involved in legal matters—but on matters that are quasi-legal, judgmental matters that involve people and situations, that you're well off. So his transfer really rounded out the secretariat, and later I found his advice absolutely invaluable.

We were forced to make one change in the secretariat prior to the end of President Nixon's first term in office. General Hershey was about to retire as director of the Selective Service System, and the White House had its eye on Curtis Tarr for his replacement. Curtis asked me how to decline. When a member of an administration, the argument that the present job is more important won't fly, nor will personal preference. I suggested he explain his distaste for selective service. Several days later he proceeded to the White House with great foreboding. After an interminable wait, the President burst out of the Oval Office, grabbed Curtis's hand and amid the handshaking thanked him profusely for accepting the job. Curtis ended up serving as the selective service director until 1972.

His replacement for reserve affairs, Richard J. Borda, was the perfect individual for the job, and we were lucky to get him. He had been vice president of personnel for the Wells Fargo Bank on the west coast. He came with strong recommendations from the business school at Stanford. He was a very well-rounded person with a lot of savoir faire.

Each assistant secretary had a counterpart on Mr. Laird's staff as well as on the Air Staff. There were deputy chiefs of the Air Staff for research and development, installation and logistics, and manpower. The only awkwardness was financial management. General Duward L. (Pete) Crow was the comptroller for the Air Force who worked directly with Bob Moot, the Department of Defense comptroller. The assistant secretary for Financial Management oversaw management systems, but had no direct control of the purse strings. On budget matters Pete Crow was Spence Schedler's boss. General Ryan and I encouraged counterparts to work closely together, and to resolve issues directly whenever possible, before they needed to come to our attention.

When I became secretary, General J.P. McConnell's four-year tour as chief of staff was about over. He accepted my appointment graciously. However, he made it clear that he was the military boss. There was one time when I wanted to become familiar with the general officers in the Air Force, particularly the three- and four-star generals, and so I asked for a book with their biographies. Word came back: "If Dr. Seamans wants to know about my senior officers, he can come and see me."

Because of McConnell's impending retirement, one of the key decisions that had to be made within a few months of my arrival was the naming of the next chief of staff. Mr. Laird was putting the bite on me for this decision, and he made it clear he wanted a younger person. An obvious candidate was George Brown, a great leader in World War II with many difficult B-17 sorties deep into Germany, McNamara's executive officer, and currently the Air Force commander in Southeast Asia. However, I felt he was still young enough to take command of system acquisition when he left Vietnam and then become Air Force chief, and even ultimately to serve as chairman of the Joint Chiefs. Fortunately for the country, this is what happened.

Harold Brown and J.P. McConnell advised me prior to February 15th that General John Ryan was their choice for the next chief of staff and that his selection would be welcomed by the senior staff. I had a lengthy conversation with General Ryan, former commander of the Strategic Air Command (SAC). I liked him, but felt I should meet with General Bruce Holloway, the present commander of SAC, before making a final determination. I went to his headquarters in Omaha, Nebraska, telling him in advance why I was coming. At the meeting I told him he was, along with one or two others, a contender for the job, so I wanted him to bear this in mind in our conversation. Then I asked him a bunch of questions about his views on Air Force matters. I got a very nice letter back saying that he appreciated being considered and the way I discussed matters with him and, of course, was sorry he wasn't selected but wanted me to know that it meant a lot to him to realize that he was considered.

General Ryan was my final choice, and I've never worked with anybody I respected more. He was direct, open, and pragmatic. No matter what the assignment, his reaction would be: "Let's not agonize, let's get on with it, let's do it." Today, looking back on my days in the Air Force, I'm proud of the team I helped put together, not only those I've discussed but the many others that space does not permit mentioning.

NASA and DOD

There were major differences between NASA and DOD. Of course DOD was still government, and a lot of its business was conducted the same way as NASA. But NASA is an independent agency and, hence, in an administrative role at NASA, I tended to work closely with many departments, directly with the Bureau of the Budget and attended meetings with the President a fair number of times. In the Department of Defense, most of my activities were within the department itself, working with the different offices of the secretary of defense and horizontally with the other services. At the Air Force, not much time was spent with departments and agencies outside DOD.

NASA had a fairly substantial international program, so I had done some traveling. But in the Department of Defense, the disposition of our forces in this country and overseas and the relationships with many foreign countries on a large scale were extremely important. In particular, Southeast Asia was front and center at that time. I enjoyed the extensive traveling, as I got to see people such as Chiang Kai-shek in Taiwan, President Chung Hee Park in Korea, and other foreign leaders. I was also privileged to stay at a U.S. residence when visiting foreign capitals and had a chance to meet many of our ambassadors.

There was a rhythm to life in the Pentagon. Upon arriving at 7:30 a.m. I was given another cup of black coffee before receiving a briefing on critical news events. I had already "read" the *Washington Post,* the *Baltimore Sun,* and the *New York Times* in the car prior to entering the building, but the defense issues were not always completely or adequately portrayed. My executive officer would be present and bring me up to date on the day's events.

I would then receive a military briefing that was similar to the one received by the chief of staff one-half hour earlier. Maps of Vietnam would show USAF activity in Vietnam and along supply routes in Laos. It was common knowledge and accepted that we were attacking the supply lines used by the North Vietnamese in Laos, but bombing of Cambodia was verboten. Hence a secret procedure was developed using the high-altitude B-52s flying toward Cambodia. The bombs were released using radio signals from the ground. The pilots could suspect but couldn't be certain whether the bombs landed in South Vietnam or Cambodia. The charts used in my briefing never showed the bombs landing in Cambodia. In my naiveté, I didn't realize until after I left the Pentagon the existence of this bombing.[2]

On Mondays, the secretary of defense held a large meeting in his conference room. Laird sat at one end of the conference table and Packard at the other with the service secretaries and the joint chiefs in between. The key staff and assistant secretary of defense observed the meeting from chairs around the perimeter of the room. On Wednesdays, I had an hour each week scheduled with Mel Laird. Dave Packard and John McLucas would also attend. These meetings were more informal, much like the sessions I had at NASA with Hugh Dryden and Jim Webb.

---

[2] Senator Symington (D-MO) telephoned me several months after I left the Pentagon to ask me if I would voluntarily testify on this subject. General Ryan confirmed at the committee hearing that I had been excluded from information about the bombing by higher authority. Senator Symington asked me if I found this upsetting. I told him it made me "damn mad," especially since I'd signed a report to Congress saying that the bombing in Cambodia was limited and inadvertent.

The development programs of NASA and DOD certainly had a common thread, but that was just part of the job in the Air Force. The procurement and maintenance of large numbers of operational aircraft and armaments provided a whole new dimension. In addition, at NASA there is no division between civilians and the military. The division creates a problem in the services because both are needed, but how do the two tie together, particularly in scientific and technical areas? It's hard to get top-grade civilian people in key positions because of the rotation of military officers in senior positions, and yet military officers must rotate in and out of nonoperational assignments. Finally, the biggest difference was the scale of the activity, measured then in financial terms $25 billion versus $6 billion at NASA (about $85 billion vs. $20 billion in today's dollars), or in personnel terms 1.3 million compared with 33,000.

However, system acquisition had much in common. The various steps included project definition, procurement planning, contractor (source) selection, contract administration, development, and operations. Each step required leadership and, during the development and operation phases, a dedicated team headed by a strong project manager.

Source Selection[3]

The Air Force weapon systems source selection process changed during my tenure. I felt that the source selection authority should review and approve the source selection plan, and then should stay away until the source evaluation board had completed its work and presented its findings and recommendations. At that point, a crisp decision usually could and should be made. It may take several weeks for clarification or to obtain additional information, but the decision should be made quickly and documented properly. In every procurement where I was the selection authority, I signed a 20- to 30-page report on the specific process, the important issues, and the rationale for the decision. But I didn't tell anybody in the building what the decision was going to be until the last day when I filled in the name of the recipient, signed it, informed the executive branch, the Congress, the contractors, and issued a press release. Up until that time,

---

[3] For comparison with NASA's source selection, see the "Long Hours, Hard Work" section in Chapter 2.

if a contractor or a congressman or senator wanted to tell me about the wonders of a company in contention, I could and would listen.

I found that previously there was a secretary's Selection Advisory Council composed of about ten members, all at about his level in the secretariat. If the F-15 aircraft source evaluation was in progress, for example, they would have gone out to Wright Field, reviewed the process, and reported back to me on the quality of the effort and the possible outcome. There would be the temptation for the advisory group to give the evaluation team advice, and certainly there was a strong likelihood for leaks and lobbying. The abolition of this committee removed another delicate area, the extent to which I kept Laird and Packard informed. Since I didn't know anything about tentative findings until the review was complete, there was no problem. I could only tell them about the plan, that the process was under way, and the probable completion date.

Many people outside the Pentagon believed that the whole evaluation process was bunkum since no matter what the Air Force determined, there would have to be a look by higher authority, and there might be political pressure for a reversal. That was never stated explicitly, and I wanted to make sure that it never happened. I felt the integrity of the process had been jeopardized in the previous administration by the award of the TFX to General Dynamics. There was no point in asking the evaluation team to do a careful job if the team members felt the decision would ultimately be based on politics.

As soon as I had reviewed the findings of the source evaluation group but before I had made a final determination in my own mind, I met with Laird and Packard alone to acquaint them with the important factors in the decision. I could discuss with them the pros and cons of the leading contractors and the probable outcome. Neither Laird nor Packard ever questioned the outcome from a political standpoint. As a matter of fact, they agreed to hold up the announcement of one award until after an election so that it couldn't be construed as assistance to a Republican senatorial candidate in whose state the award was going to be made.

Contracting

The C-5 contract was conceived by the Air Force with the best of intentions. The contract provided Lockheed with specific performance objectives measured in terms of payload, range, speed, and short field capability. The plane also required many special features, including the capability of "kneeling" to take loads on and off. Kneeling involved developing special rugged hydraulics for the landing gear to lift the plane up and down so that the floor of a truck or a loading platform would be the same height as the floor of the airplane's cargo bay. Thus a truck could back straight up to the C-5 to move its cargo easily, or the C-5 could squat, lower its ramp, and move cargo directly onto the ground.

But the contract tied many of these requirements to corporate profit and loss in such a way that the priority for these capabilities was no longer in Air Force hands. In addition, the contract attempted to price the production runs and the spare parts prior to the first development flights. McNamara and Assistant Secretary of the Air Force Robert H. Charles introduced the concept of a total procurement package to increase contractor cost-effectiveness. Military contractors had often submitted intentionally low bids for a research and development contract, reasoning that they could recoup financial losses by padding the production contract. Under the total procurement package concept, the Air Force awarded a contract for a complete weapon system, including research and development, test, and production of the system. The package included specific schedules, cost ceilings, and financial incentives. The C-5 program was the first to use this complicated arrangement. The impact of these and other contractual arrangements is discussed in the subsequent section "Galaxy and the Company Books."

Funds are authorized and appropriated for spending by the Congress for specific purposes and given amounts. No form of contracting can abrogate this responsibility; it can only be assumed by the specified government agency, in this case the Air Force. For this reason I believe in hands-on contract administration by a competent government team with appropriate oversight by the general management of the agency. The contract should state the objectives clearly and offer rewards for excellent performance and penalties when objectives are not achieved. The "carrot and stick" approach may be based simply on cost when the development is essentially complete, but when design uncertainties still exist, an award fee may be preferable. In this way, corporate managers will feel it in their self-interest to assign top-quality personnel to the undertaking.

Project Manager

It is sometimes said that legal documents don't determine the outcome of a procurement, people do. I believe both are important. The project team must operate within an understandable, helpful framework. Team members must be competent and nobody more so than the project manager. The Air Force conducted a study in 1969 of how long project managers were serving before reassignment. The more senior members were serving less than three years. The data was a little biased because some of the members included in the survey were still on the job and hadn't completed their tour of duty. But we definitely needed to extend the time spent by the project team before reassignment. As a result of this study, the Air Force made a conscious effort to keep senior people on these projects long enough that they felt responsible. Avoidance of mistakes, an early transfer, and a clean record should not be the modus operandi. I don't really believe many officers had that point of view. However, large projects require a real commitment from the project leadership. A real commitment requires staying on the job long enough so that the fruits of the effort become apparent.

There had been sufficient difficulty with both the C-5 and the F-111 aircraft that General Ryan and the Air Staff recognized that changes had to be made in the acquisition process. They fully supported the need for experienced project managers on extended assignments.

Project Reviews

Prior to becoming secretary of the Air Force, I had been pretty thoroughly exposed to the ballistic missile program since these missiles had also served NASA as launch vehicles. I also had worked with the Air Force on the development of the Manned Orbiting Laboratory[4] (MOL) since this project made use of the Gemini manned space capsule. Since Mel Laird told me I was selected for the job because of my

---

[4] About a month after my arrival, I started picking up rumors of the MOL demise. I asked Mel Laird for a day in court if such was under consideration. My day arrived on a Saturday afternoon when General Stewart and I met with Laird, Kissinger, and President Nixon in the Oval Office. Kissinger called on Monday to tell me I had made an excellent presentation, and on Tuesday MOL was canceled.

technical background, I was particularly anxious to attend a project status review to look at these and other programs. My predecessor had scheduled these reviews on a quarterly basis, and so at the appointed time I left my office and crossed the long Pentagon corridor to the designated conference room. Imagine my surprise when just inside the door a colonel saluted me, clicked his heels, and announced in a loud voice, "The secretary of the Air Force." Everybody jumped to attention and stood at attention until I sat in my designated position at the end of a large conference table.

Now NASA people are no slouches at putting on presentations but what unfolded was obviously polished and rehearsed to the nth degree. One of the questions I remember asking that first day was: "To what extent has this material been reviewed prior to the time that it is being presented here?" The answer was nineteen times. Nineteen levels in the organization had taken a crack at it! You can imagine that there wasn't much left that was controversial. That was the last quarterly review and the last review that was massaged nineteen times during my watch.

I believe a month is the right period of time to review important programs. That's what I knew worked in NASA. That's how financial books are kept and how corporate directors run their business. After the reordering of status reviews, the assistant secretaries and the deputy chiefs became the board of directors, with either the secretary, under secretary, chief, or vice chief presiding. The status of the twelve to fifteen most critical projects was presented monthly, and actions were taken on the spot as appropriate. If the decision impacted on a particular field commander, he would be given the opportunity to participate. The project manager couldn't be expected to be present monthly because of his busy schedule, but he was required to attend at least quarterly and more frequently during critical periods. His deputy or someone from the Air Staff would represent the project at interim meetings.

The following eight projects were either so important to the future of the Air Force or so much in the public that they along with seven others were on the critical list for the secretary's monthly reviews. Another twenty-five or so were reviewed by Grant Hansen and his counterpart, Lt. Gen. Otto Glasser. Hundreds of others were delegated to elements of the Systems and Logistics Commands in different parts of the country.

Galaxy and the Company Books

I had worked with Raymond L. Bisplinghoff at MIT, and he later directed NASA's aeronautical and space research and was a member of the Air Force's Scientific Advisory Board. So he was a natural person to call on when we were in technical trouble with the C-5 aircraft. We knew we had electronic and landing gear problems, but we also knew the most serious issues related to the wing structure, and that was Bisplinghoff's specialty.

He and his committee reported that the overall performance was very close to the specifications. The only real deviation from specifications was one caused by structural fatigue, leading to a reduction in life expectancy. The plane could fly at the maximum gross weight specified and could take off from the prescribed runways. However, every heavy load lessened the aircraft's life expectancy. The full load could only be carried when it was really important, as, for example, in the 1973 airlift to Israel.[5] In that mission the aircraft were each carrying a couple of tanks, each weighing over 100,000 pounds. That was pretty close to design. Most of the time when the flying was to keep up crew proficiency and the like, the plane was not fully loaded and life expectancy loss was minimized.

By keeping track of the load and the turbulence of the air on each flight as well as careful frequent inspection of the wing structure, 20,000 hours of flying rather than the 30,000 specified could be obtained. So the Air Force ended up nursing its 80 heavy load cargo planes along by tail number[6] prior to eventually redesigning and rebuilding the wings. It wasn't cheap, or easy, but the Air Force did it.

The Galaxy (C-5) contract not only had profit incentives for a variety of performance factors, but also had the production prices tied to cost. In the original plan, there were five aircraft for development followed by fifty-seven in Run A and a comparable number in Run B. In January 1969, during the remaining weeks of the Johnson

---

[5] Golda Meir, the Israeli prime minister at the time, was at the airport to greet the first Galaxy. When the plane's nose lifted and two tanks emerged, she reportedly was deeply moved and bent over to kiss the ground.

[6] For both logistical and operational purposes, aircraft are identified by the number normally painted on the tail.

administration, the contract called for procurement of long lead items for Run B. This was agreed to, but at a later time, Lockheed claimed that the Air Force had triggered the procurement of the entire Run B. The Air Force ultimately procured eighty aircraft, not the originally planned 120, so that this number became still another item for future negotiation.

By the summer of 1970, the Air Force estimated that Lockheed was in arrears by $200 to $400 million. Of course, Lockheed didn't agree, but to continue, there needed to be an understanding with them, including a renegotiation of the contract. But what was Lockheed's financial capability? I've always felt that the Air Force shouldn't have to take responsibility for the solvency of a contractor. This is what I mean by looking at the corporate books: What degree of liquidity does the contractor have? What degree of flexibility? Is the company having difficulty borrowing from the banks? These kinds of questions should not be the government's responsibility.

The difficulty was that at Mr. Packard's level, the government was no longer looking just at Lockheed's work on the C-5, but also on the Navy's Polaris missile, the National Reconnaissance Office's classified satellite projects, and an advanced helicopter for the Army. Packard had to look at the totality of Lockheed's defense business. But the only other major business they had was the commercial L-1011 airplane, with its engines coming from the then-bankrupt Rolls Royce, and a good production run of C-130 transports. So when Dave Packard met with Dan Houghton, the CEO of Lockheed, I'm certain all the cards were on the table. From the Air Force's standpoint, Dave negotiated a cost contract for the C-5 with a loss to Lockheed of $200 million. The Air Force was finally back in control of the Galaxy, and the resulting fleet of aircraft has been invaluable.

From TFX to a Successful F-111

The experimental fighter designated the TFX (Tactical Fighter, Experimental) was in trouble almost before it started in the early 1960s. Under the rubric of cost saving, McNamara decreed that the Air Force and the Navy would develop a new fighter in concert. The Air Force, as the lead service, conducted a source selection clearly favoring the Boeing Aircraft Company. Gene Zuckert, the Air Force secretary, spent several hours reviewing this finding with McNamara and returned to announce that General Dynamics was the winner. A series of congressional committees reviewed the decision; the most acrimonious occurred when McNamara appeared before Senator John McClellan from Arkansas. The battle lines were drawn and continued until Laird assumed office, when the tide turned. I remember a luncheon in Laird's office attended by Laird, Senator McClellan, General McConnell, and myself. The conversation went smoothly, with General McConnell and the Senator swapping stories about their early years growing up in Arkansas. By dessert, Senator McClellan was almost in tears as he expressed thanks for being invited into the Pentagon for the first time in eight years. He went on to advise us he was no longer an adversary of the TFX, now becoming the F-111, the nation's first variable wing aircraft.

The F-111 had two serious technical problems, the carry-through structure supporting the wings and the advanced avionics for the F-111D. In order to swing the wings aft as the speed increased, the wings were pivoted on a single box-like structure, made of high tensile steel. Steel was used to obtain strength and save weight to preserve the capability for carrier landings that were never to take place. Unfortunately, several aircraft were lost when this structure failed. The failure was ultimately traced to small fractures within the steel that grew with time. The science of fracture mechanics was in its infancy, as was the technology for finding these hairline cracks. It was known that the probability of failure was greatest at low temperature when the steel became more brittle. The Air Force was faced with grounding the fleet unless a reliable ground test could be found. The situation was worsened by the fact that the Australians had bet their future defense on the F-111s, which they had been sold at a bargain basement price.

The airplane was designed for 7g pullup maneuvers and 2g's nose down. A fixture was developed by General Dynamics to apply the equivalent forces on the wings, tail, and fuselage using hydraulic jacks. After cold soaking the aircraft at −40 degrees Fahrenheit for 24 hours, all aircraft in the inventory and all subsequent planes off the assembly were put through a strenuous regimen of simulated attack maneuvers. There were only a handful of test failures and no subsequent accidents in flight.

In order to provide all-weather capability, the F-111Ds were provided with a television-like monitor, which blended radar signals with prestored mapping information. Unbelievably, the scope had 1,600 lines, four times the number in our home TV sets. General Dynamics had contracted with North American for this equipment on a fixed-price contract. They in turn had contracted with United Aircraft and Hughes for critical components. These contractors were millions of dollars in the hole and refused to continue. The Air Force had 96 F-111Ds ready to fly except for this missing equipment, and General Dynamics blamed the Air Force, claiming it had forced them to use this untested, unproven equipment.

Until this horrible experience, I hadn't realized how cash payments could be used to motivate production contractors. Normally, partial payments of 80 percent are made on a weekly or monthly basis as aircraft leave the production line. The remainder is paid after the aircraft are successfully tested in flight. General Dynamics was receiving 90 percent progress payments on a daily basis. I couldn't believe it. The improvement in their cash flow was enormous thanks to Air Force largesse. I called Dave Lewis, the CEO of General Dynamics, to advise him the Air Force was shifting to normal progress payments and why. Poor performance on the advanced avionics was one of the reasons I cited. His first response was unprintable, but henceforth General Dynamics knuckled down, and the F-111 became a wonderful addition to the Air Force stable of aircraft.

The F-15 — Can the Air Force Do It Right?

One of the keys to a successful F-15 aircraft project was the appointment of a truly competent manager and to agreement on his availability long enough for development and early production. Major General Benjamin Bellis was the answer. He had successfully managed the government's side of the SR-71, the highly successful strategic reconnaissance aircraft that flew Mach 3 at 70,000 feet. Kelly Johnson of Lockheed was his industrial counterpart.

The Air Force had need of a new air superiority fighter—a plane that could protect air space over a battle zone. At the start there were two principal issues: (1) the requirements and (2) the method of contracting. The USAF fought hard for a single purpose aircraft and for a contract with simple incentives. The people in Defense Research and Engineering did not agree. That office felt the F-15 should be capable of close support of troops and wanted a total package arrangement as used for the C-5. The military was fully behind a true superiority fighter. Ultimately, the only compromise made in the design was the placement of hard points under the wing for possible attachment of bombs. We had little support within the Pentagon for our method of contracting. The C-5 contracting had not as yet been discredited. Finally in desperation, I went to Dave Packard with a paper outlining the type of contract we wanted. He approved the paper much to the dismay of his staff. Procurement could then commence.

Many requirements have to be addressed prior to the request for proposals (RFP). I'll only mention a few. In order to minimize cost, the maximum speed was to be limited to Mach 2.2, thereby permitting an all-aluminum airplane. Higher speeds cause excessive heating for aluminum, thereby requiring titanium or equally expensive alloys. Next, it was decided the F-15 would be designed to carry only one person (the pilot) in order to keep the weight down and improve maneuverability. This decision placed great emphasis on the electronic package, including the radar. Detection and tracking of targets had to be accomplished semi-automatically to avoid pilot overload. And the radar had to operate at long range and all altitudes, which required pulse doppler to avoid radar ground clutter, especially when flying low.

The final lineup of contractors included McDonnell Douglas for the aircraft and its electronics, but with a "fly before you buy" between Hughes Aircraft and Westinghouse for the radar. The engines (two per aircraft) were provided by Pratt & Whitney.

After contractor selection, we instituted a "scrub down" to see if the cost of either development or unit production could be reduced. The procurement officers were distressed, believing it unfair and risky to contract for a design different from the original. General Bellis reviewed the cost-saving changes at a meeting with General Ryan, General William Momyer (chief of the Tactical Air Command), and myself. After we made a favorable determination on these changes, we briefed Mr. Packard and the staff of Defense Research and Engineering, who thought the outcome was great. We were starting to rebuild confidence in the Air Force.

General Bellis was responsible not only for the aircraft but for the Pratt & Whitney engine that was going to be used in both the F-15 and also in the Navy's F-14. This was a tough assignment. He had to contend with somewhat different Navy regulations. Before going ahead with the

F-15 production, the engine had to pass a 150-hour endurance run under strictly controlled conditions. In mid-summer when he had reached 135 hours, it became known he had omitted the Mach 2.2 tests at 40,000 feet. A brouhaha of major consequence ensued within the Pentagon and on Capitol Hill. General Bellis had used good judgment in eliminating this one point for fear of a fire, but bad judgment in not coming clean at an Air Force status review. If he had, we could have informed Packard, the Congress, and the press. But as a result we had to restart the endurance run and nearly had to forfeit our basic contract.

Another close call had occurred previously, just after the McDonnell Douglas Company was selected to build the aircraft. I was attending a commanders conference at the SAC base in Puerto Rico when I received a phone message from John McLucas. He said Laird would ask us to rebid the F-15 contract unless we inserted a proper affirmative action clause in the present contract by the end of the day. Somehow the lack of a proper clause had been omitted and had become public knowledge, and Father Theodore Hesburgh, president of Notre Dame and a member of the President's affirmative action committee, had castigated the Department of Defense for its oversight. I suggested John straighten it out. He said, "Mel insists you go to St. Louis this afternoon." I went back to the conference and announced, "I must be in St. Louis in the shortest possible time." I was soon taken by SAC police, sirens screaming, to the flight line where a tanker was off-loading fuel as we approached. The tanker was on alert, prepared to take off with such a heavy load that it would crash if one engine failed. My departure was not such an emergency. A ladder descended from the nose, and a sergeant grabbed my bags from General Holloway and up the ladder I went.

By mid-afternoon we taxied onto the ramp at the McDonnell Douglas plant. Awaiting us was a small coterie around Mr. McDonnell, CEO of the company, and a TV newsman was standing nearby. After shaking hands all around, I vented my pique about the lack of an affirmative action plan in front of the TV, and we then headed for the negotiation table. The contract was appropriately amended in a few hours, Mel Laird was appeased, and I returned to Puerto Rico.

As of this writing over 1,200 F-15s have come off the McDonnell Douglas assembly line. The airplane has performed well, a testimonial to General Bellis and his team.

AWACS, from Air Defense to Battle Control

This project wasn't planned; it just seemed to grow on its own. AWACS stands for Airborne Warning and Control System. Its purpose originally was to track incoming Soviet bombers coming across Canada so that they could be intercepted before reaching and destroying U.S. targets. The key to success was obviously not the aircraft but the electronics. Contracts were let to Hughes Aircraft and Westinghouse to develop experimental radar prototypes. By the time the prototypes were ready for flight test, interest in air defense was rapidly waning. Since we and the Soviets had ballistic missiles that could reach most parts of our respective countries in less than one-half hour, why worry about lumbering bombers?

It appeared to some in the Air Force that there was an important role for AWACS in the European theater. With the ability to track hundreds of airplanes, couldn't AWACS be used for air battle control? The enthusiasm for this idea was just sufficient within Defense Research and Engineering to maintain prototype funding of the radar. Westinghouse won the flyoff and Boeing was selected to provide the carrier. The slowly turning 30-foot rotodome mounted atop the fuselage gave a space-like appearance to the converted 707-type aircraft.

General David C. Jones was given the responsibility to take the first airborne prototype to Europe for testing. He determined that it was not only possible to keep track of large numbers of friendly and potential enemy aircraft, but transponders could be placed on Army ground units so that the position of friendly fighters could be related to the ground war. General Jones returned from Europe greatly enthusiastic, only to find many skeptics at home. Among the issues that remained was AWACS vulnerability. Could the AWACS remain useful while far from the battle area?

While decisions on AWACS were still in question, I had a chance to demonstrate its capabilities to Ken Rush, the new deputy secretary of defense. We took off from Andrews Air Force Base outside of Washington and flew a racetrack course above the District of Columbia at 35,000 feet. At this altitude, we were simultaneously tracking 450 commercial planes located all the way from New York to North Carolina.

Twenty years later, AWACS was in the Air Force inventory and proved to be most valuable in coordinating air activities during Operation Desert Storm. The Air Force still hasn't demonstrated the effectiveness of AWACS in more intense engagements, but hopefully the need will never arise.

AX—The Tension of Roles and Missions

The development of the AX further validated the advantages of prototype procurement, and it also provided a good lesson in "roles and missions." Experience in Southeast Asia showed that both the Air Force and the Army needed to improve their close air support of ground forces. Air Force pilots flying high-speed aircraft were sometimes having difficulty locating much less hitting ground targets. The Air Force was thought to be ineffective. The Army was finding helicopters invaluable for transportation and rescue but had such limited firepower they couldn't destroy enemy outposts. The Army had a helicopter gunship under development that promised true combat firepower. But what about helicopter vulnerability in a battle zone? The Air Force suspected the Army of underestimating helicopter attrition rates and thus wanted hard data. When the Army response appeared favorable—that is, helicopters apparently could withstand battle damage—the Air Force claimed that the Army kept helicopters in their operational inventory if they could retrieve any identifiable part of the aircraft. In the vernacular, they would "keep the tail number if they could retrieve the wheels." In short, interservice rivalry was intense.

Two service secretaries entered this arena where angels should fear to tread. I met with my counterpart, Stan Resor, secretary of the Army. We agreed that a fixed-wing, close-support aircraft, the AX, was probably needed, and that the Air Force should be responsible for its development. A simple memorandum of understanding to this effect was signed by both parties.

The result was truly remarkable. General Ryan wanted me to rescind the memo by claiming I didn't understand the import of my actions. Stan Resor received similar recrimination from General William Westmoreland, the Army chief of staff, who claimed he was giving away the store.

I later heard that Stan had said, "Bob and I must have done something right. Both of our staffs told us that we had sold them down the river." Ryan was so upset, he sent the vice chief, General John Meyers, to see me as his emissary. He was afraid he would be too emotional if he came himself. I attempted to explain my action. The Air Force was complaining about the Army's role, but had little say in the matter because it didn't have the effective tools of the trade. The Air Force began the procurement with little comfort on either side. In addition, Mr. Laird reportedly felt that if the Air Force developed a successful close-support aircraft, it would probably be transferred to the Army.

The requirements for the AX included ease of maintenance in the field, durability, a steel tub around the cockpit for pilot protection against small arms, and a 30-mm Gatling gun the size of a Volkswagen in the nose. There were two strong contenders, Northrop and Fairchild Hiller, and they were both provided funds to build two prototypes. Northrop built the A-9, and Fairchild handled the A-10.

Carefully selected pilots were given the job of flying and scoring the two airplanes. Fairchild not only won the flyoff, but their aircraft with two outboard engines was given high marks for maintainability. The announcement of the winner drew criticism on many fronts. Senator Lowell Weicker from Connecticut, the home of Lycombing, the engine manufacturer for the A-9 aircraft, was quick to point out that the outcome must have been influenced by Governor Nelson Rockefeller of New York and Vice President Spiro Agnew since Fairchild was located on Long Island and outside Washington in Maryland.

The plane was not sleek and beautiful—it earned its name the Warthog for good reason. More than 700 aircraft were produced, some of which saw service in Operation Desert Storm. Questions have been raised about the A-10's vulnerability and effectiveness. The role of the Air Force in close ground engagements is still an open issue.

Interdiction—The Mice Win

The Air Force had a variety of combat missions in Southeast Asia. General Abrams could designate an area each day for pattern bombing using B-52s at high altitude. F-4s and F-105s were used for reconnaissance, close support, strafing, and precision bombing of designated targets. The precision was much improved as the war progressed by the use of "smart" bombs. The most effective technique relied on a laser designator that illuminated the target and a traditional bomb with a laser homing device attached to its nose and movable fins attached to the tail. The laser signal is projected onto the target from another aircraft or from the attacking aircraft itself. The homing device is locked on the signal and then the bomb can be released. The results were particularly effective on railroad bridges across rivers in North Vietnam.

But the most difficult Air Force assignment was the interdiction of supplies traveling down the Ho Chi Minh trails in the Laotian panhandle. The difficulty was caused by the unusual geography. Prior to World War II, Indochina was a French colony that included what is now Vietnam, Laos, and Cambodia. The boundaries of these nations were determined on the basis of race, culture, topography, and history. Vietnam held all the shore line along the South China Sea. The middle section of Vietnam backed up against the Laotian mountains is in some places only 30 miles wide and is about 250 miles long. In order to effect a cease-fire in 1954, the French agreed to a division between the North and the South along the 17th parallel. This boundary was later expanded to form a demilitarized zone usually called the DMZ. Even after the cease-fire, the Viet Cong (VC) continued their insurrection in the South, and the North considered it in their interest to aid and abet the VC with a wide variety of military supplies.

The supplies were delivered to North Vietnam by rail from China and into Haiphong Harbor by Soviet ships.[7] The resourcefulness of the Vietnamese moving supplies south cannot be overemphasized. When their rail lines along the coast were temporarily destroyed, they would enter the Laotian panhandle further north. Laos borders Thailand along the Mekong River, and its panhandle is 75 to 125 miles wide. The region is mountainous, with foliage and countless rivers and streams. The supplies moved south on Soviet trucks, and when the roads were under repair, on the back of animals and men. As the war in South Vietnam involved local harassment by the VC, not synchronized warfare, delivery wasn't usually time critical. The major battle at Khe Sanh was an exception.

Several technologies were used to stem the flow of supplies. The previous administration made a heavy investment in the development and placing of sensors along the myriad of trails. These took the form of probes that could be air-dropped and would penetrate the earth just deep enough that the antenna would remain above ground. The antennae were designed to

---

[7] The North Vietnamese also used Sihanoukville in Cambodia as a port of entry for supplies distributed primarily in the Mekong Delta area south of Saigon.

appear as small trees or bushes. There were a variety of sensors, including seismic, to measure the passage of trucks, people, and animals. Microphones were used to pick up the sounds of motor vehicles or even conversation. Signals from the sensors were transmitted to a control center just across the Mekong River in Nakhon Phanom, Thailand. It was hoped to build up sufficient knowledge of supply routes and their frequency of use that aircraft could be dispatched to areas of activity in real time.

On occasion an audio device would be detected, and the excited conversation of the Vietnamese could be heard until the unit was destroyed. A particularly intriguing episode occurred when the parachute slowing the descent of a probe caught in a tree. The excitement of the Vietnamese rapidly increased, and then one of them could be heard climbing the tree and reaching for the device. There was suddenly a loud noise of splintering timber and a shriek, but the final outcome will forever remain in doubt; the sensor went off the air.

An adjunct to the Air Force and Navy fighter planes used to impede the flow of supplies was the gunship, which was introduced into Southeast Asia by the previous administration in 1965. A large opening was cut in the side of C-47s so that Gatling miniguns could be fired at the ground traffic as the C-47 circled overhead. By 1969 the Vietnam traffic flow was mostly at night, and the gunships were becoming more sophisticated, using both C-119 and C-130 aircraft.[8] The development was strictly at the grass roots level, with Major Ron Terry at Wright Field, Ohio, in the lead. He was receiving little support from the Tactical Air Command. I called a meeting to discuss the lack of support with General Momyer and General Ryan. Momyer argued the gunship would not be effective in Europe because of its vulnerability. So why spend resources on a special mission not often encountered? I argued for increased effectiveness in an existing war, with the thought that Vietnam insurgency might not be unique in the future.

Ron Terry received grudging assistance. Ultimately, he installed 105-mm cannons in C-130 aircraft. A wide variety of infrared detectors and displays were used for locating trucks and aiming the guns.

---

[8] The Air Force names for the gunships were "Spooky," "Stinger," and "Spectre" when using, respectively, the C-47, C-119, and C-130 aircraft.

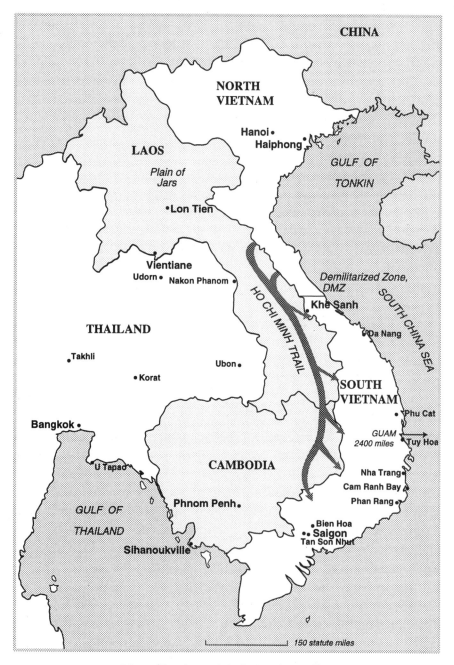

Map of Southeast Asia during the conflict.
U.S. airbases are indicated in the smallest type.

When a truck filled with weapons was hit, the explosion registered clearly on the infrared screen. For an extended period trucks were being tracked down and destroyed at a rate of 150 to 200 per night. It is estimated that, by the spring of 1972, 24,500 trucks had been destroyed. Suddenly, the trucks disappeared. We never determined whether the Vietnamese concealed roads, used other forms of transportation, or reduced the flow of supplies. It was a cat-and-mouse game, and the mice apparently won.

The B-1 — Big League Competition

General McConnell was anxious to get the development of the B-1 bomber under way, so he made an appointment with Packard to discuss the issue. I came to the meeting not too well-grounded in the subject but recognizing that the B-52s were twenty to twenty-five years old and would need to be replaced in a decade or so. I also felt that preliminary design studies such as those on AMSA (Advanced Manned Strategic Aircraft) were expensive and a slick way to avoid decisions. At the meeting, McConnell agreed to cancel further production of the planned fleet of 250 FB-111s. The bomber version of the F-111 had limited load capacity and range and was not a true strategic bomber. It was agreed to initiate plans for a new strategic bomber, the B-1.

Missiles can be land based and delivered with great accuracy, or aboard submarines hidden under the sea and launched with precision depending primarily on the submarine's navigational accuracy. However, missiles, unlike a bomber, cannot be recalled once launched. And missiles also lack the flexibility of carrying a wide variety of small or large warheads. So bombers continue to be part of the U.S. inventory in the age of missiles, constituting the third leg of the so-called Triad (the other two legs are the land-based and submarine-based nuclear missiles).

Before a bomber reaches its target, it must penetrate the nation's land mass, which in the case of the Soviet Union was heavily fortified. Simulated tests against our own defenses showed the advantages of high speed and low altitude, preferably under 200 feet. Cost projections favored high subsonic speeds rather than supersonic by a wide margin (Mach 0.95 compared with Mach 1.1). Another large cost factor was the use of a swing wing. Although more complicated, this configuration reduced size and power requirements. Consideration was given to a titanium, rather than a steel, carry-through structure to support the wing pivots. However, the weight saving didn't justify the sharp increase in cost.

Our experience with the F-111 was clearly beneficial to the design of the larger swing wing aircraft. Like the F-111, supersonic speeds would be obtainable at high altitude although this was not a basic requirement. Two big issues remained: the avionics and the procedure for bailing out prior to an imminent crash. I took the lead on the electronics, with Jack Ryan on the escape system. Our overall objective was to provide a bomber that would be sufficiently low cost to be accepted by the public and with performance suitable for the Strategic Air Command, in the 1980s and beyond.

Every new airplane is going to have equipment, such as avionics, added later. It's a mistake to add all conceivable equipment at the start and run up excessive costs due to development difficulties and ensuing delays. So we tried to keep the avionics simple. We thought we had made good decisions on the avionics, and then we started to get disturbing signals. We were trying to keep the unit price down to $4 million. The avionics team came in and said, "We can't guarantee it, but we think we're up to at least $8 million."

We just had to take a "meat axe" approach. One of the requirements resulted from the need to fly long distances "on the deck" (at low altitude) with sufficient precision that the radar could determine position along a coastline. If the plane flew at an altitude of only 200 feet until near the coast and then climbed to 400 feet, it made a tremendous difference in the navigation error that could be allowed and still distinguish landmarks along a coastline with reasonable probability. By relaxing the requirement, we no longer needed new inertial navigation systems; rather, we could use commercial equipment. The lessening of this requirement saved lots of dollars.

Another cost-cutting example was the B-1 escape capsule. One of the initial requirements was the ability to escape at very high speed and low altitude even from a tumbling aircraft. The escape capsule became much like a reentry vehicle but stabilized aerodynamically with spoilers, flaps, and brakes. General Ryan had to decide whether to continue with the capsule or change to a conventional bail-out procedure using individual parachutes. He concluded that there would seldom be the need to eject at 720 miles per hour (Mach 0.95) "on the deck." By changing the requirement to being able to eject at speeds less than 450 mph, the capsule became much simpler and cheaper to build. The cost was reduced, and from a development standpoint the capsule was no longer "the long pole in the tent." The B-1 development was canceled during the Carter administration in favor of the B-2 stealth bomber. Its stealth permits it to hide from enemy radar and hence penetrate and approach targets without alerting ground defenses. Even more recently, 100 B-1s were produced and added to the strategic inventory in the Reagan years.

A highly classified visit to the Strategic Air Command in Omaha, Nebraska, could reveal the SIOP (Single Integrated Operational Plan). This plan spelled out each target in the Soviet Union that would be attacked in the event of a nuclear war. The plan designated the tonnage, the specific delivery missile or bomber, and the routes the bombers would follow. Each missile wing, submarine, and bomber would be provided its specific assignments in the event of a Soviet attack. It is truly awesome to contemplate the damage that would have been inflicted on both countries following a nuclear exchange.

I found there was a much less frightening tit for tat between U.S. and Soviet air attachés. When the Bolshoi Ballet came to the Kennedy Center, the Soviet air attaché in Washington sent me front row tickets. Since I was going to be traveling on that date, I asked to have the tickets returned with a note of explanation and thanks. Just prior to this, the pregnant wife of one of our attachés in Moscow had apparently been jostled at a Soviet reception. So my invitation was returned to the attaché in a willfully insulting way. A junior officer slammed the tickets on his desk and said, "The secretary wouldn't consider accepting your invitation."

When I found out this had happened, I discussed the stupidity with Jack Ryan, who agreed to help put a stop to the nonsense. Each year the U.S. Air Force took the foreign air attachés to an active USAF base. That year the attachés were going to SAC headquarters. Jack Ryan and I decided to include this particular Soviet attaché, a first. The air attachés weren't shown much more than would have been seen by a Rotary Club, but the Soviet attaché was ecstatic. He was gloating because none of his superiors had ever had such an opportunity.

DSS—The Unthinkable Alert

The U.S. and the Soviets held each other at bay for the 50-year cold war with "credible deterrents." The deterrent was based on another cold war concept, "mutually assured destruction." The Soviets had to believe that no matter how massively they attacked us, their destruction was assured. No matter how or when we were attacked, sufficient weapons would be delivered on their military and industrial centers that they could no longer function as a viable country. Our ability to counterattack hinged on reliable early warning.

A series of systems was erected to provide such warning. The first included radars and communication stations located along northern Alaska, Canada, and Greenland and was called appropriately the DEW line for Distant Early Warning. This system could give adequate warning against incoming bombers but had no capability to alert the U.S. of a ballistic missile attack. In the late 1950s, a large-scale effort was made to install BMEWS (Ballistic Missile Early Warning System). Large tracking radars were installed in Alaska, Greenland, and Scotland to detect Soviet ballistic missiles before they could reach the continental United States. This system led to one disquieting moment that served to place SAC on alert. Fortunately, the USA wasn't under attack; rather, the Moon was rising above the horizon. As the Soviet buildup continued with further construction of missile silos and increasing numbers of submarines stationed near our shores, deficiencies in BMEWS became obvious. The Defense Support Satellite (DSS) was built to fill this gap in our strategic defense.

DSS truly relied on advanced technology and for this reason gave me the most concern of all weapons systems under development. However, I needn't have worried. Aerojet General did an exceptionally fine job with the sensors, although security was so tight there was little public recognition. DSS was placed in a synchronous orbit in a position from which it could observe all rocket launches as they appeared above the Soviet atmosphere. By analysis of the gas-burning plumes behind the rockets, DSS could discriminate between known types of rockets. DSS also had the capability to count the number of rockets and to determine the destination of their payloads minute by minute. This knowledge would permit the President to be whisked by helicopter from the White House to the Airborne Command Post on the runway at Andrews Air Force Base, ready for takeoff. When aloft, the Command Authority was much more secure than at the previous post located inside the mountain near Camp David.

The Airborne Command Post was fashioned from a Boeing 747. Information is received aboard from DSS, BMEWS, and all other available sources. Computers and electronic maps give the President real-time information on which U.S. targets are to be attacked. It is common knowledge that the "football" (a briefcase containing the launch codes for nuclear weapons) goes wherever the President goes. Using the "football" the President can order a nuclear counterattack by communications to the commands specified for this purpose. When aboard the Command Post, redundant channels are available for such dispatches. A nuclear war would be such a world disaster it's almost unthinkable. But only by such thinking could we continue to have a credible deterrent during the cold war.

Campus Life

With the Vietnam War on, it was a trying time to be secretary of the Air Force. With American campuses in an uproar over the war, it was probably even more difficult to be a student.

Our son Joe was in the Harvard class of 1970. In his junior year, there were antiwar riots in Harvard Square and in the Yard. I was out west visiting the Air Force Academy during one such happening. When I called Gene, who was in our Brattle Street house at the time, she said, "We've got some pretty bad things going on here. Joe came to supper to tell me about it."

"How does Joe feel about it?," I asked.

She said, "He thinks it's all nonsense."

"Well, I'm glad of that."

The next morning at about 5:30, the phone rang in my room at the Air Force Academy. It was my brother Peter, who said, "I just want to warn you as early as possible that when the police stormed into University Hall after midnight to break up the demonstrations, Joe was one of the students there."

"Well, Peter, I talked to Gene last night, and Joe was with her, saying the whole thing is nuts."

"That may be," Peter said, "but on the front page of the *Boston Globe* this morning, there's a picture of Joe catching a young lady jumping out of University Hall. The caption refers to Joe Seamans, son of Secretary of the Air Force Seamans, and says he was at University Hall when the police came in."

It turned out that Joe had gone to the movies with some friends. Afterwards, seeing the bright lights in Harvard Yard, they had gone over to investigate. They found a guy with a bullhorn rallying anti-war sentiment, inviting fellow protesters to sit in at University Hall. Joe and his friends, obviously curious, went inside. At about three in the morning, the protesters got word over the radio that police were arriving at dawn. Joe said it was all sort of exciting until he saw real policemen coming up the stairs of University Hall-all of them appearing to be about six feet, six inches and dressed in riot gear. He found a simple solution: jump out the window. It was quite a drop, and he landed pretty hard. He saw a girl who had already jumped who appeared to have broken her ankle. He helped her, then tried to catch another girl jumping. Down she came, her skirts up around her head, and flash! went the camera. The reporter came up and asked Joe's name.

Joe said, "I don't want to get my father in any trouble." Of course, that did it. If he had said, "I'm Joe Smith," the paper never would have used the picture.

Before I went to Cambridge on the weekend, I said to Gene, "I want to talk to Joe." When I arrived, I put on some old clothes and bicycled all around Harvard Yard to see what was really going on. There were posters that looked as though they had been printed by Russian revolutionaries in 1917. When I finally met up with Joe, he said, "All of this is really confusing. It's so hard to figure out what's going on. I can't possibly study in my room at night. Somebody comes in asking me to sign a protocol, and I do so just to get them out of the room." His roommate was a member of Students for a Democratic Society (SDS). He said, "I go over to the Radcliffe library if I want to get anything done."

Then he went on to say, "I do wonder about the Cabots. You know, they own a lot of the land around the Harvard Medical School, and as a result they're making an awful lot of money supporting the school's land-acquisition process."

"First of all," I said, "it's not the same Cabot. There are two Cabots. Cabot, Cabot, and Forbes are the ones in the real estate business. Tom Cabot, a very distant cousin, has done an awful lot for the Harvard Medical School, but he and his family are not getting anything out of the real estate. You know Tom Cabot. You grew up with his grandson. That charge is totally inaccurate."

Prior to Harvard, Joe had gone to Phillips Academy in Andover, where he was captain of the crew and became interested in cinematography. He not only got the art prize but also received a prize for his photography. While at Harvard, he took a number of courses in photography at MIT with the great Minor White. By the time he graduated cum laude in English, he knew he wanted to go into television.

He had trouble getting a job. After about two months he came down to Washington with his friend Andy Schlesinger, son of historian Arthur M. Schlesinger, Jr., to visit me. The two of them were picked up at the airport by my aide, Colonel Mike Cook, and arrived at my office looking dreadful! They had long hair, of course, and their clothes were a bit shabby. The military people who saw them were surprised by the lack of a Seamans dress code!

At Harvard Joe had met a young woman from Wellesley named Elizabeth ("Betsy") Nadas, who was a year ahead of him in school. Her father was the leading children's cardiologist in the country. Betsy had a job working with *Mr. Rogers' Neighborhood,* the public television program produced at WQED in Pittsburgh. She wrote scripts for the show and played the role of Mrs. McFeely, the postman's wife. Betsy wrote Joe and said, "Why don't you come out here and go to work for Mr. Rogers? You probably won't make much money, but you'll have some fun." Joe worked for the program for three months without seeing a paycheck—pushing crates around and doing whatever needed doing-then finally went on the payroll. A few years after his arrival in Pittsburgh, he and Betsy were married at the Memorial Chapel in Harvard Yard. They now have two children. After leaving the show for a while, Betsy is back working part-time with Mr. Rogers, helping with a great variety of programs.

Joe's first big filmmaking opportunity came in 1976. As part of America's bicentennial, the National Geographical Society financed a film about *Hokule'a,* a catamaran that sailed from Hawaii to Tahiti and navigated entirely by the traditional Polynesian method. Joe was supposed to be an assistant only, helping with pre- and postproduction. But when his boss, the cinematographer, boarded the ship and set out with the Hawaiians for Tahiti, he became immediately seasick. Before the ship had left the Hawaiian Islands, Joe had replaced him as cinematographer. The footage he took was so good that the resulting program was made thirty minutes longer than originally budgeted.

May followed Kathy to Dobbs. By the time she got there, she was faced with the same sort of uproar that her older brother Joe was having to deal with at Harvard. Coming home to Washington was no escape. There were all kinds of characters coming into town every Saturday and Sunday. Back at Dobbs, she didn't always hew perfectly to the line. In time, she fell in love with Gene Baldwin, who worked at the school. When we went to May's graduation in 1971, she told us she planned to marry. Plans already had been made for May to spend that summer in Mexico. We hoped that she would think over her decision, but it didn't work out that way.

May and Gene were married on August 27, 1971, while visiting the Baldwins. They had a daughter in 1978, but were divorced two years later. May came to Cambridge to live and married a very talented dentist, Elliot Kronstein, on October 11, 1986.

May has been trained as a hospice worker, and she helped set up a hospice for people suffering from AIDS. It has provided a wonderful final refuge for hundreds of people. She is licensed as a nurse's assistant and a home health aide. She is a gifted healer.

Globetrotting

When you're involved in the Air Force, flying is *de rigueur*. I did a lot of flying, visiting Air Force installations around the globe. The Air Force has many more bases than NASA had centers, so there was a lot of ground to cover. I made it a point in my four-plus years as secretary to visit nearly every major installation inside and outside the United States, and Gene accompanied me on about one-third of the journeys.

In the continental United States, or *in conus*, as it is called, I flew in a Jetstar, a high-performance airplane. It had four jet engines and made very good time—a far cry from the Gulfstreams used at NASA. Outside this country we made use of a KC-135, initially designed for use as a tanker to refuel bombers. It had no windows in the fuselage. If passengers wanted to see what was going on, they had to sit between the pilot and the copilot. The belly of the plane, normally unfurnished, was cleverly outfitted with insertable modules for executive travel. One module contained tripledecker bunks, where sixteen could sleep on a long flight; the module behind that had four tables, each of which sat four for dining and conferences; and behind that were the galley and lavatory modules.

The military let me know that I ought to get to Southeast Asia as soon as I could. I took the first of many trips in May 1969. On a typical trip, it was a seven- or eight-hour flight from Andrews Air Force Base to Anchorage, but flying across five time zones, we got there before lunch. At that time there were a lot of small bases all over Alaska, particularly on the northern perimeter, as well as very large radars around Fairbanks, all of which kept track of what the Soviets were doing. In my different trips west, I stopped long enough in Alaska to visit many of them. From Alaska we flew across the Pacific to Tokyo, arriving at one of the three major U.S. Air Force bases near that city—often in time for dinner! From Washington to Alaska to Japan, the Sun never set.

On my Asian trips, we proceeded to Saigon via various routes. Sometimes we stopped in Okinawa, sometimes in Taiwan; sometimes we went through the Philippines. Quite a few times I went to South Korea, where we had many bases. I was lectured by the president of South Korea on the stupidity of the United States protecting Japan against Soviet aggression. He argued that the U.S. and South Korea were putting up major resources, while the Japanese weren't putting up a nickel for their own defense. Instead, they were using the money to compete economically against us. I didn't have a very good answer for him, except to say that the United States didn't particularly like the idea of having the Japanese rearm.

"Then why did you let the Germans?," he asked.

While in South Korea we, like many foreigners, were invited to a kee sing party. We went to a restaurant, where every guest had a lovely young woman assigned to him. While we had drinks, our respective ladies were obliged to keep our glasses full. When we sat down on our haunches around the dining table, our ladies fed us. There was dancing with music; then at the prescribed time, perhaps ten o'clock, a whistle went off and the ladies left. Seemingly, they couldn't wait to get out of there!

On one remarkable trip we went into Taipei for a meeting with Chiang Kai-shek. He was an old man at the time, and his wife was ailing, so I did not meet her. We went for tea. I thought the meeting would last half an hour. Instead, we stayed two. Chiang had a lot of questions, then gave me a lecture. Sometimes he didn't agree with the translator's rendering of his Chinese into English, so he corrected him in English. When we were leaving, I said that I hoped to visit the museum which housed the art treasures taken out of mainland China. I had heard that it housed so much, it could have had a different exhibit every day for a year. It was past museum closing time, but Chiang said, "Don't worry, the museum will be open for you." I was given the grand tour.

I don't know how many trips I took from Andrews Air Force Base to Tan Son Nhut Air Base, just north of Saigon. Once when Gene came with me, we stayed in what was known as the Little White House inside Saigon. During our two days together in Vietnam, she visited (always with a flak jacket in her car) an orphanage, a "typical" home of a civil employee, a Vietnamese army hospital, and a rural school, where she was unexpectedly called on to make a speech. The street scenes were unforgettable to her, with their endless piles of trash and numerous little houses shingled with metal from beer cans. On that exhausting trip, we made fourteen stops in two weeks, going as far as central Australia. When we got home, Gene had a severe sore throat and had lost her voice and tummy, though not her head!

Other times I stayed at Tan Son Nhut, a huge base with more traffic flying in and out than is found at any major United States civilian airport. Our military presence there was massive. After a briefing with the commander of U.S. forces in Southeast Asia, General Creighton Abrams, and a meeting with our ambassador, Ellsworth Bunker, I flew up-country, where the Air Force had another seven major bases. One base in Da Nang was used for training the South Vietnamese—what was known in those days as Vietnamization. Tens of thousands were taught English, then taught to be mechanics, pilots, and navigators. Goodness knows what has happened to them all; they were well trained.

We flew pretty close to the DMZ, or demilitarized zone, to take a look at the classified communications outpost on Monkey Mountain. Then we flew into Thailand and the big base at U-Tapoa, where the Air Force had large numbers of B-52 bombers. Fifty or sixty B-52s flew missions every night, each with over 100 bombs in its bays underneath. Then we flew on to Bangkok and visited two big up-country bases along the Mekong River. In later trips, I visited Laos, a never-never land if ever there was one. G. McMurtrie Godley, our ambassador, was in charge of all of the American interests there, a fact that was supposed to be sub rosa, though it was known to everybody. Godley was the person who had almost bought our house in Washington in June 1968, before the deal fell through. It wasn't until I was sitting in his living room in Vientiane that I found out what the problem had been. Our real estate agent had been unwilling to give his agents their share of the commission. Once before leaving Washington, I had received a back-channel communiqué that Godley's wife, who was Greek, was having a terrible time getting kitty litter for her pets. When I arrived at the front door of the embassy in Vientiane, my Air Force jeep was loaded with a 100-pound package of kitty litter tied with a great big red bow.

Godley took me on tours in his small twin-engine plane. Once we flew into the very mountainous northern region of Laos. The peaks are not that high—maybe 9,000 feet on average—but they are very steep and numerous, seemingly placed at random. The weather was terrible, and I finally convinced Godley that, much as I wanted to see Lon Tien, where the Hmong forces were centered, I could probably get along without it!

On my next trip to Laos, we did land at Lon Tien, and I was glad we had stayed aloft on the previous visit. The runway slopes upward toward a mountain, which precludes a go-round if the first landing is missed. When taking off, planes taxi to the high mountain end of the runway and take off going downhill in the opposite direction.

General Van Poa, head of the Hmong, met us in his jeep and drove to a receiving line of men and women dressed in ceremonial garb. There are five subtribes of Hmong, and Van Poa had a wife from each to avoid hurt feelings. I never found out whether they were in the receiving line.

Before lunch Ambassador Godley and I were greeted by all the

Hmong present in a unique ceremony. We sat on our haunches, and each guest passed by on all fours, providing us with an egg and a jigger of white lightning. Then each tied a good-luck string around our wrists. Fortunately, after several Hmongs had gone by, Godley whispered to me, "You don't have to eat and drink it all!"

After lunch, we went to Plaine des Jarres (Plain of Jars), where the Hmong had been fighting the North Vietnamese for years. The jars were everywhere, man-sized and carved out of stone. Nobody knows how these ancient relics came to be there. We also saw fourteen-year-old boys on the battlefield carrying rifles taller than themselves.

Several days later, I attended the king of Thailand's annual garden party for the diplomatic corps. Ambassador Leonard Unger asked if I would like to attend and if so to bring my white tie and tails. The garden was in the middle of the palace grounds. Each diplomatic team was lined up by seniority, with no more than two guests standing behind each ambassador and his wife. As the king and queen proceeded along the line, each ambassador bowed and his wife attempted a low curtsey.

Later the king and queen separated, and introductions took place. I suggested to Len Unger that we visit with the beautiful queen first. When introduced, I explained my connection with MIT and said how happy we were to have her daughter studying there. Her face darkened, as she told me that her daughter had forsaken her heritage and would never return to Thailand. Changing the subject, I discussed my trip to Laos and the strings still around my wrist. The queen explained that when she returned from Laos the strings went as high as her elbow. I decided I was not prepared to deal with royalty.

Of course, much of the Air Force's activities involved NATO, and Gene and I made several trips to Europe, ranging from Norway to Turkey, with many stops along our routes. An unusual trip taken without Gene started at our bomber base in Loring, Maine, on the Canadian border. Morale was excellent there even though the greenhead flies took visible chunks of skin from the unsuspecting. The next stop was Iceland, where we shared the Keflavík Airport with commercial aviation. As we were just below the Arctic Circle, the days were long, and service people were having great difficulty putting their children to bed.

Our northern route then took us up the west coast of Greenland, stopping at Sandrestrom at the end of a long, scenic fiord. From there north to Thule the weather was worsening. The spectacular coastline and ice-filled ocean became less and less visible as we approached the airport, and there was also a strong crosswind. The pilot explained that, in landing, we would follow our glide slope until we reached an altitude of 200 feet. If at that point we didn't see the strobes guiding us to the end of the runway, we would go back. After a successful landing in "zero-zero" conditions (zero visibility), it took great skill for the pilot to taxi to the terminal on the icy runways. The base commander couldn't believe we were able to land. The base was in a "class-four" whiteout, in which not even ground vehicles are allowed to travel.

Such conditions can last several days, but fortunately for us the weather cleared later that evening and we went on a tour of the various BMEWS facilities. A large 500-foot fixed antenna is situated on the promontory that faces north towards potential incoming missiles. Looking north the eye could see nothing but a sea of ice and snow.

From Thule, we flew directly to Fairbanks, Alaska, in just three-and-a-half hours. This part of the trip was of particular interest to Bob Mateson, a Washington friend originally from Minnesota. In the course of many summers he had canoed the northernmost rivers in the hemisphere—from the Hudson Bay, west across Canada, over the divide, and down the Yukon into Fairbanks. I wanted his views on how the Air Force was handling environmental issues. As a consequence of my inspection with Bob, the Air Force removed and sold for scrap over two million fuel drums that had accumulated at its bases in Alaska.

On this trip, we inspected BMEWS facilities near Fairbanks that were similar to the ones at Thule. We then flew to a fighter base in Galena on the banks of the Yukon. Bob Mateson was especially anxious to meet a particular Indian guide who was famous for his knowledge of the region. During the two-hour period when Bob was with him, the river rose six to eight feet around the guide's house. Bob was advised not to canoe on the lower Yukon, as silt can fill clothing and drown a person in minutes. However, he was undeterred and made the trip the following summer. After touring Alaska, we flew home by way of Minot Air Force Base in North Dakota, where there is a major ICBM (intercontinental ballistic missile) installation.

Gene and I took another memorable trip, to South America. First stop was Panama, where the South American command is located. We met the Army people who ran the canal, and Gene had a chance to operate a lock as a ship was passing through. Then we flew on to Caracas, Venezuela, where we stayed with the ambassador. This was the first time we were made aware of terrorism firsthand. When we went to a local club for some tennis, we were escorted by a station wagon in front of us and another just behind, both carrying heavily armed bodyguards. Guns could be seen sticking out of the two vehicles in all directions. While we played, there were men on all sides of the court holding drawn guns. Our ambassador told us that he always carried cyanide pills, in the event he was taken hostage.

From there we flew to Rio de Janeiro, Brasília, and Buenos Aires, where a general strike was in progress. We could not get a car through to our embassy, so Ambassador John Lodge (brother of Henry Cabot Lodge, Jr.) came over to our hotel to chat with us. The president of the country kindly loaned us his plane so that we could fly to Cordova, location of the training facility for Argentinean pilots. The next day the Lodges gave us a very elegant dinner party. Ambassador Lodge, who was so handsome he could have been a movie actor, had a beautiful voice and loved to sing at parties. After he sang for us, he unexpectedly called on Gene. She stood up in front of that group of seventy-five jeweled strangers, with her foot in a cast and the musicians poised. Then, God bless her, no words came out! (She later admitted to stage fright.) My military assistant, Colonel Cook, who also had a very nice voice, immediately jumped up and sang "Hello, Dolly!" with her, saving the day. From Buenos Aires we traveled to Lima, Peru, where the Peruvian government flew us to join a National Geographic trip to Cuzco and Machu Picchu.

Another time we went together to Mexico City. Gene and I knew the ambassador, Bob McBride, a personal friend of the president of Mexico. The president finally said he would see me, but only as a friend of the ambassador, not in my capacity as the secretary of the Air Force. The McBrides took us over to call on him. As it was customary to exchange gifts on such occasions, Gene had brought along a hooked rug and a quilted pillow from a store in Georgetown, Massachusetts—crafts she especially enjoyed giving on our various foreign trips. She had written a description and history of them and, before a given trip, had it translated into the languages of the countries we would be visiting. When the president and his wife opened their package, things changed. All of a sudden they were very open and friendly, offering to show us the presidential mansion and, a Mexican custom, the family shrine in their basement honoring their dead.

The POWs Come Home

One of the biggest heartaches during my term as secretary of the Air Force was knowing that around 700 American flyers were incarcerated in Hanoi and were being treated badly. Yet the POW (prisoner of war) crisis also resulted in one of my most satisfying Air Force experiences, one I wouldn't have fully enjoyed if I had resigned when Elliot Richardson and I first discussed the matter.

Our family took a ski vacation to Vail in 1970. There were nine of us in the party, which meant there was always an odd person out when we paired up for the two-passenger chairlifts. Once when I was the odd person, I paired up by accident with a handsome woman named Joan Pollard. She asked me what I did. I said I worked in Washington.

"As what?"

"Secretary."

"Secretary of what?"

"Of the Air Force."

By the time we had reached the top, she had told me that her husband, Ben, was an Air Force navigator, that he had been captured by the North Vietnamese six years before, and that she was head of the Colorado POW-MIA (missing in action) families. Joan Pollard was a very interesting person, and getting to know a POW wife brought the tragedy even closer to home. We corresponded, and when she came to Washington on POW-MIA business, I invited her to my office. We were telling the families everything we could, but it was nice for me, and I hope for her, to have this personal contact.

Finally the great day came—February 14, 1973—when the Air Force was allowed to send C-141s into Hanoi to pick up our prisoners of war. For each planeload there was a manifest, which was relayed to me as soon as a plane left the ground in Hanoi. I checked list after list without finding Ben's name. Then on the very last manifest-there it was! I called Joan in Colorado and said very emotionally, "Ben has left Hanoi."

A week or two after I had left the Air Force, a large White House dinner was held for the POWs and service people involved in their release. John McLucas, the new secretary of the Air Force, felt very strongly that I should be invited, but H. R. Haldemann was absolutely adamant that I not be included. Though I would have been very gratified to share in the moment, considering all that came later I consider Haldemann's rejection a great compliment. We have stayed in touch with the Pollards and greatly admire not only Ben's conduct in prison—where he somehow managed secretly to build a slide rule and teach engineering to his fellow inmates—but also his adjustment to freedom and his renewed family life.

Leaving the Air Force

By law, the secretaries of the Army, Navy, and Air Force are in charge of managing their departments—overseeing budgets, development and procurement of equipment, training of forces, and so on. They are not, however, in the chain of command on military operations, which passes from the President through the secretary of defense and the Joint Chiefs of Staff to the forces in the field. So while I shared in the responsibility, I was not directly involved in decisions on national policy or the conduct of the war.

However, I could and did make my views known. In my second year in the Air Force, I was invited to the White House to discuss the Pentagon budget with the President, Henry Kissinger, Mel Laird, and the other service secretaries. Resor, Chafee, and I were given six minutes each to make presentations. The Air Force, being youngest of the armed services, always goes last; so I waited while Resor and Chafee talked. Their discussions concerned morale and a perceived need for some sort of White House ceremony to honor our military heroes.

I felt that, with the country groaning out loud over Vietnam, we ought to be able to have a candid discussion about what was really going on. I said at the meeting that it ought to be recognized at the policy level that we were paying a heavy price for what was happening in Southeast Asia and that our national security was being jeopardized. The simple way of looking at it, I said, was that we were providing the

South Vietnamese people with all kinds of equipment, which we would not have available to us in the event of a flare-up elsewhere in the world. Perhaps more important, I said, was the difficulty we were having recruiting good people. ROTC (Reserve Officers' Training Corps) programs were being canceled at our best universities.

When I was through, I got a twenty-five-minute sermon from the President on the domino theory and other theoretical justifications for the war. He ended by saying that if I didn't feel I could handle the job of secretary of the Air Force, there were plenty of other people who could.

After two years at the Air Force, I considered resigning. For me, it was a question of how much longer I wanted to participate, how much longer I felt I could contribute. While skiing at Vail, I drafted a letter, which I read to our assembled family. It wasn't a letter of resignation. Instead, it said that unless America made every effort to disengage itself from Vietnam at the earliest possible date, I would resign. I gave the letter to Laird. He liked the fact that I was willing to take a stand and told me so. In fact, he told me that he was putting what weight he could behind the notion of an early and speedy disengagement.

In 1972, just before Nixon's reelection, Robert N. Ginsburg, the two-star general in charge of Air Force public affairs, felt that both General Jack Ryan (the chief of staff) and I should individually meet "off the record" with key members of the media. We invited some leading members of the press to a cocktail-dinner party at Ginsburg's house. After dinner, we sat around in Bob's living room, and they fired questions at me.

At about eleven o'clock, someone said he had just one more question before leaving. What was the chance that, if we were to meet here three years later, the war would still be on? If I had had any sense, I would have said, "You may be here, but I won't." Instead, I recapitulated the reasons why I thought we would be pulling out soon—Vietnamization, the peace negotiations, and so on. Then I added, "If you ask me, 'Is there any *possible chance* we'll be there three years from now,' I'd have to say, 'Yes, there is that possibility.' "

On my way home in the car, the radio news announced that a high-ranking civilian in the Air Force said there was a good chance America would be in Southeast Asia another three years. Subsequent stories embellished this one. By the time the next issue of *Time* magazine came out, it stated categorically that Robert C. Seamans, Jr., secretary of the Air Force, predicted we would be in Southeast Asia for three years! I was on the White House blacklist thereafter.

About three days after Nixon's reelection, all presidential appointees within the Defense Department were called into Laird's office. He told us we were all receiving identical letters from the President, as were political appointees in every department and agency. He had the letter read to us. It asked for our resignations, so that the President would have the flexibility he needed to make his second term as effective as possible. Laird then advised us all to give him a signed note, tendering our resignations. He said he would then call the White House and say he had the resignations in hand. "Then," he told us, "when you really want to resign, write me an appropriate letter."

Laird himself resigned shortly afterward. I was en route to Southeast Asia and Antarctica[9] when I received a communiqué saying that Elliot Richardson had just been appointed to replace Laird and that he wanted to speak with me. I called him and told him my itinerary for the next two weeks. I told him I would turn around and come back if he needed to speak with me immediately.

He said, "No, that sounds like a great trip. By all means, go ahead with it. But," he added, "what's up between you and Haldemann?"[10]

"What do you mean?"

"Well, I find that the White House, in effect, wants your head."

"As a matter of fact," I told him, "I do want to get out of the Air Force."

"Well," he said, "I've heard that you have an opportunity to head up the Sloan-Kettering Cancer Institute."

I told him that it wasn't definite but that I had been talking with Lawrence Rockefeller about the possibility.

When I returned from my trip, the Sloan-Kettering opportunity did not pan out. I told Elliot that I did not have an immediate need to resign. But early in 1973, I wrote him, saying that I wished to resign from the

---

[9] Guy Stever and I met at the Air Force facility in Christ church, New Zealand. There we and our party were fitted out with the necessary boots, parkas, and other gear for the Antarctic. The flight to McMurdo Bay aboard a C-141 took four-and-a-half hours. While there we visited a penguin rookery, the South Pole station, the Soviet base at Vostok, and the New Zealand station, while flying in a C-130 equipped with skis.

[10] Nixon's chief of staff H.R. Haldeman.

Air Force in May. I had just been nominated president of the National Academy of Engineering (NAE), which was holding its annual meeting in Washington during the first week in May, and this seemed a good time to make the transition. Ironically, Elliot Richardson left the Defense Department before I did, to become U.S. attorney general.

As I prepared to leave the Air Force, I thought of all the talented and dedicated military and civilian individuals with whom I'd worked. I attempted to say good-bye while offering thanks by visiting several commands during my last week in office. I went to the headquarters of Strategic Air, Tactical Air, Airlift, Logistics, and Systems Acquisition Commands. Flying from one base to the next, I thought of our accomplishments and failures.

The most perplexing was our role in Southeast Asia. Our job there was finished and that was a blessing, but what had been accomplished? Could South Vietnam continue on its own without U.S. military involvement? Even so, was it worth the many American and Vietnamese lives? It was difficult even then to realize the magnitude of the casualties sustained during eight years of fighting. For an extended period, more than 300 body bags were filled and returned to the United States every week. Of course the public was greatly upset.

I remembered the rioting in Chicago at the time of the Democratic Convention in 1968, the shooting at Kent State University, the burning of American flags, and the trashing of ROTC classrooms. I believe there is general agreement now that our military intervention in Southeast Asia was a mistake. There has been a loss of confidence in our leaders. The public is much more skeptical of government than prior to our involvement in Asia. How did this happen?

Before addressing this question directly, it should be kept in mind that both the Johnson and Nixon administrations were faced with the deadly serious cold war. They had to consider the possible impact of their actions on responses by the Chinese and the Soviets in unison or separately. In addition, President Nixon didn't want to upset his delicate plans for opening relations with China. With that said, let me make a few personal observations with the benefit of twenty-twenty hindsight.

The McNamara-Johnson idea was to keep raising the ante. They reasoned that at some point, the communists would break. On the contrary, the more the communists were pounded, the greater their resolve became.

Johnson and McNamara were also naive in their belief that they could hold back communism by fighting the Viet Cong and the North Vietnamese army primarily in the South while attempting to close the supply routes through Laos.

It was the Nixon-Kissinger view that we should withdraw from Southeast Asia, but only by first strengthening South Vietnam so that it could go it alone. I believed in and supported Vietnamization, as this policy was called. However, this policy by itself didn't permit withdrawal rapidly enough. U.S. casualties during the Nixon administration (21,000) amounted to 36 percent of Americans killed during the war.

The only way to have prevented a communist takeover was to conduct an operation similar to Desert Storm, the successful thwarting of Iraq's plans to take over Kuwait in 1992. We had to attack the North Vietnamese forces and their supporting infrastructure at home. We did take such action but much too late and only when the North Vietnamese reneged on the 1972 peace accords. Although highly controversial at home, the so-called Christmas bombing of selected targets around Hanoi and the mining of Haiphong Harbor immediately brought the North Vietnamese back to the negotiating table. Early in the war, shipments along the rail lines from China to Hanoi could have been the Air Force's number one assignment, and Haiphong Harbor could have been blockaded by the Navy. Intercepting supplies and weapons in transit through Laos was nearly impossible, however. Moreover, trying to do so placed no direct pressure on North Vietnam.

If the risks of more direct attack on North Vietnam were felt to be too great, the only other option was to negotiate a U.S. withdrawal. President Nixon felt that South Vietnam would fall if the United States withdrew and that the whole area would be subject to a communist takeover according to the domino theory. Furthermore, he felt that the United States would be judged an unreliable ally in the world arena. I can only comment that this perception of the Untied States would, even if it existed, be temporary, but our extended stay in Southeast Asia was a permanent loss at home.

Today, the United States is the only superpower. We do have broad responsibilities in our own interests as well as in the interests of the world community. That said, it is my strong belief that we should never again risk the lives of thousands of our young men and women unless the future of our country is directly at stake.

CHAPTER FOUR

Reentry

The National Academy of Engineering

THE National Academy of Engineering (NAE) was founded in 1964 as a spinoff from the National Academy of Sciences (NAS). There had been quite a bit of concern within the engineering community that the NAS, founded by President Lincoln during the Civil War, did not adequately represent engineering on the Washington scene. So a small group of engineers within the NAS got together and recommended that a corresponding academy of engineering be formed. The NAS did not want the engineers approaching Congress for a new charter because the old one was very favorable and no one wanted to see it reopened for revision. So the engineers said, "Okay, as long as we can select our own members and can participate in the management of the National Research Council [the "business end" of the NAS, responsible for carrying out governmental studies], that will be fine."

But it wasn't fine. The third president of the NAE, Clarence Linder, and Philip Handler, president of the NAS, didn't speak. They were absolutely at loggerheads. Linder felt that the engineers weren't getting the time of day, and Handler got sick and tired of being hassled by the engineers. So the engineers came to the conclusion that they had to separate from the scientists, lock, stock, and barrel. It was onto this scene that I came as fourth president of the academy in May 1973.

The morning before the formal annual meeting convened I met with the academy's council. It was agreed that the membership should be asked at the annual meeting to abrogate the agreement with the NAS and go off on their own. Up to this point, I knew little about the rift between the two groups. I had been a member since 1967, and I had met with Clarence Linder a couple of times before becoming president to try to find out what he thought the academy's function and goals were. But all I got out of him was a diatribe against Phil Handler.

At the annual meeting, the membership balked at the split. It was finally agreed that no vote would be taken until the new president had had a chance to review the situation. I hadn't been there very long when I started receiving a large number of letters. Out of maybe a hundred, only five or six were for secession. The common feeling seemed to be that there was great strength in the alliance between the scientific community and the engineering community. Furthermore, the two form a continuum—from pure science (the theory of relativity, for example) all the way through the application of science and "engineering science" to commercial development and production. To split the two ends of the continuum would be artificial and do a great disservice to the country. The Congress wouldn't always know on a given issue which academy to turn to. On top of which, our academy didn't have more than $50,000 in the bank, whereas the NAS had an endowment of $50 million!

During the summer I composed a two-page letter to Phil Handler in which I not only said that I didn't think we ought to split but also gave a formula for working more closely together. I pulled the old bureaucratic trick that I had learned at NASA and the Air Force of going in to see him with a draft letter and asking, "How would you like to receive this letter?"

He read it and said, "But it's too late."

I said, "Phil, if you believe it's the right thing to do and I believe it's the right thing to do, then it's not too late."

"Well, send me the letter." After he had received it, he invited me to come to the summer meeting of the council of the NAS and explain myself. Some members of the science academy were as negative toward the engineers as some of the engineers were toward them. Nevertheless, I told them that I thought it was a mistake to split. Furthermore, I said that the NAS, being the senior organization, had the final responsibility on all matters of policy dealing with the federal government. Reaffirming that we engineers still wanted a say in the National Research Council, I suggested that from thenceforth the council have seven members from the NAS, four members from the NAE, and two members from the Institute of Medicine. That way, if there was a real rift, the scientists would always have a majority and could swing any vote their way. I further suggested that Phil Handler be chairman of the council, that I be vice chairman, and that all government studies, including those already under way in the NAE, be transferred to the National Research Council.

The scientists voted to back my proposals. Now all I had to do was get my fellow engineers to agree! Some weren't all that enthusiastic, but the majority went along with me. And Phil Handler and I came to enjoy being with each other and working together. I found him a brilliant, versatile man.

ERDA

At the fall 1973 council meeting of the National Academy of Engineering, one of our members, W. Kenneth Davis from Bechtel, said, "From where I sit, I think this country is facing a difficult energy situation." In 1970, for the first time, the United States had imported more oil than it had exported. Oil imports were growing rapidly every year. I responded to Davis's concern by saying that the NAE ought to investigate. I invited him to chair a committee, and he agreed. By sheer coincidence, the committee met for the first time in October 1973, one week after the Arab oil embargo had begun. Cars were standing in long lines for partial fill-ups. Suddenly energy was the hot topic. Our committee's report, published in the spring of the following year, got a lot of publicity. We had a press conference and testified before Congress.

When Congress started looking at what we ought to do about the energy situation, they began talking about putting together an agency like NASA. Much to my surprise, I was invited to the Executive Office to discuss personnel for such an undertaking. After meeting with several supernumeraries, I was introduced to Frank Zarb, associate director of the Office of Management and Budget (OMB). He told me I was on the shortlist to be the administrator of a new energy agency. "Frank," I said, "you've got to be kidding." The legislation for the proposed new agency hadn't been passed yet, and I doubted that it would be. But I said I would be happy to discuss it with the President, if and when it became a reality.

In the summer of 1974, President Nixon resigned and was replaced by Gerald Ford. The enabling legislation for the Energy

Research and Development Agency (ERDA) was passed in early fall, and I got a call from Frank Zarb. He said they were going to take the entire Atomic Energy Commission (AEC) and move it in under the umbrella of the new agency. Then he told me the President was sending my name up to the Hill as his nominee for first administrator of ERDA. I reminded him that I had said I would consider the appointment after a talk with the President. The next morning I went in to see President Ford. When I entered the Oval Office, the President stood up from his desk, came over, shook hands, and offered me a cup of coffee. The whole atmosphere of the Ford White House was much more open and relaxed than that under Nixon.

I told President Ford that I had five concerns I wanted addressed before I accepted his nomination. Most importantly, I asked that I be given the opportunity to recommend the seven other presidential appointees who would be working with me. I also wanted to be sure that he would not insist on a particular appointee if I disapproved of the person. We shook hands on this and the other matters of concern to me, and in the following days I was appointed by the President. A month later I was confirmed by the Senate.

Even before I was confirmed I was deluged with phone calls and letters from people who wanted jobs, who had programs to sell, or who wanted interviews for the media. Inside the government there were also many demands. The energy budget had to be presented to Congress in two months; the seven presidential appointees had to be selected; and, most important, arrangements had to be made for placing existing agencies under the ERDA umbrella.

Dixie Lee Ray chaired the AEC, and she pressed hard for transfer of the AEC immediately. Several of the commissioners had already left, so it was next to impossible for her to make policy decisions. In addition, she said the uncertainties associated with the transfer were causing severe morale problems. Dixie was a colorful character. Unmarried, she lived in a trailer with her two dogs, who inhabited her office during working hours. The smaller dog enjoyed jumping into a visitor's lap, and the larger one would sniff from behind when least expected.

Guy Stever, a friend and former associate at MIT, was director of the National Science Foundation (NSF) and President Ford's science advisor. While I was awaiting confirmation, he offered me an office where I could have all the required secretarial amenities. However, the Executive Office felt it would be presumptive to move into a government office until confirmed by the Senate. So I was left stranded and shorthanded in my NAE office. Unfortunately, security in the National Academy of Sciences and particularly around my office was porous. It would have been easy for the curious to enter my office or even my files. So I kept all sensitive information in my briefcase, and I kept my briefcase with me.

At this critical juncture Hugh Loweth—deputy associate director for science, energy, and space in the OMB—came to my rescue. He assigned Ray Walters,[1] an energetic, talented young bureaucrat to work with me full-time. Together, we located an office building that could be renovated and occupied within six weeks. Admittedly on December 29, the day I was sworn in, there was still an uninstalled toilet parked in my office and awaiting a plumber. My office looked out on the railroad tracks in southeast Washington, and every time a train went by, conversation would stop and pictures on the wall would have to be realigned.

Meanwhile, the selection of key individuals was the most time-consuming and critical of the early problems I faced. The seven presidential appointments included my deputy and heads of the six branches of ERDA prescribed by Congress: Alternate Fuels, including solar and fusion; Nuclear, including uranium enrichment; Fossil Fuels; National Security, including nuclear weapons; Environment; and Conservation. Key staff functions also requiring careful screening were the general counsel, public affairs, congressional affairs, international affairs, administration, comptroller, operations manager for fifty field installations, and the secretary of the general advisory committee. For some of these positions, the White House gave me lists of thirty to forty names, many of which were unsuitable.

I was, of course, desperate for a deputy who could help share my responsibilities. Previous government experience, capability, and availability were key ingredients for the job. I was fortunate to locate Robert Fri, who had been deputy to William D. Ruckelshaus in the Environmental Protection Agency (EPA) and was currently an executive in the management consulting firm of McKinsey and Company. He was

---

[1] Prior to Ray Walters, I was provided with an individual who told me I should "wire" the agency by placing moles in each center of activity. He lasted two days.

nominated by the President before year's end and was working full-time by late January.

There were excellent people already working in the Atomic Energy Commission and in the transferred sections of the Department of the Interior, EPA, and the National Science Foundation. These groups provided a resource from which many jobs were filled. Ultimately our key people came from these and other government agencies, along with five from industry.

I remember clearly recruiting Phillip White on the day before Christmas. In Beverly Farms for a few days, I ran up a significant phone bill. Phil was vice president for research at Standard Oil of Indiana. He was on a White House list and had the right professional credentials. After making suitable reference checks, I was able to reach Phil to discuss his appointment as assistant administrator for Fossil Fuels. He ultimately accepted and provided ERDA with valuable industrial experience. I finished phoning on December 24, just in time to complete my Christmas shopping.

The following May, Dan announced one morning, "I'm not going back to St. Albans next year. If I spend another year at St. Albans, we're going to hate each other for the rest of our lives." I was never certain why. We told him we wished he had mentioned this a bit earlier, but we did look at other schools. We finally all agreed and sent Dan to the Putney School in Vermont. It was a great success. I think that's where Dan "found himself." He hiked for hours in the surrounding woods and learned how to use tools and to take dents out of cars (a useful skill for a teenage driver), as well as all the usual high school subjects. He also grew to be six feet, four inches. By his senior year, he was attracted by the idea of taking time off before college, but we encouraged him to continue his schooling without interruption, at least until he had a year or two of college under his belt. He chose the University of California at Santa Cruz, which is known for its strong emphasis on undergraduate education and its beautiful setting. Dan traveled to California alone, camped out on the campus, and got himself accepted for that fall.

Santa Cruz also had a good music department. All during his adolescence, Dan had been almost obsessed with music. At our Washington house on Idaho avenue, a great group of friends that Dan had organized used to come over and play rock and roll together in our basement. I had a pair of Air Force mechanic's earmuffs I could wear in my study when they were practicing.

After college, Dan continued living on the west coast and became a professional musician, playing the double bass. He eventually moved to Oakland, sharing various houses with fellow musicians. They would advertise for a female housemate, "to avoid the locker-room ambiance that can easily prevail in an all-male household." A young lady from Rochester, Minnesota, named Linda Hill answered one of the ads. After an interview, she was accepted, but before she moved in she was offered less expensive accommodations elsewhere. Over a period of a year or so, Linda and Dan saw more and more of each other, and on October 30, 1988, my seventieth birthday, they were married. After several years in California, Dan and Linda gravitated eastward to Manhattan, to Brooklyn, and finally to Vermont. Now the parents of two children, they have since returned to California.

ERDA provided many unique experiences, of which I'll briefly mention four: inspecting the nuclear navy with Admiral Hyman Rickover; an unexpected disembarkation from *Viva* at night for a White House meeting the following morning; inspecting nuclear facilities in the Soviet Union with my counterpart, Commissioner Petrosyant; and visiting briefly with the Shah of Iran.

Admiral Rickover

On paper, Hyman Rickover, the "father of America's nuclear navy," reported to the secretary of the Navy and the administrator of the Energy Research and Development Agency. He was a dour, wiry individual who spoke his mind forcefully. When ERDA was formed, he had been on active duty for fifty-six years. He retired seven years later, in 1981. He prided himself on his spartan lifestyle, and Congress loved him. I went with him to most of the installations over which he had jurisdiction.

Most memorable was a twenty-hour period spent with the admiral aboard the attack submarine *Cravallus*. We met at the General Dynamics plant in Groton, Connecticut. We first entered the building where General Dynamics was designing a Trident submarine. The design teams were located in a wooden mockup of the vessel. Every new element or redesign was translated into a wooden part, in order to check compatibility and access. Rickover told me that his man at General Dynamics and the plant manager each had to provide him with weekly reports. If there was nothing significant they could say so, but if this happened several weeks in a row, he would know they were lying and would immediately inspect. Of course, he would also zoom in on any differences between the government and industry reports.

We left the dock about midnight and headed for deep water beyond the continental shelf. Rickover was not present for breakfast in the small officers' wardroom. He was conducting his daily hour of calisthenics. Once he finished exercising, he took me on a detailed tour ending in an aft cabin where warrant officers control the reactor, propulsion, and ship's electrical power. The skipper was ordered by Rickover to proceed at flank speed. Suddenly, Rickover gave the command for full reverse. The vessel shivered and shook and became hard to handle at near-zero speed. Finally the skipper blinked and said, "Ahead quarter-speed." The Admiral said, "I didn't give that order. Change the watch for a repeat."

Later I was on the conning tower as we entered Long Island Sound at high speed. It was late afternoon in February, and I was thoroughly enjoying the ride. I was reflecting on Rickover's tremendous success developing the nuclear navy, but wondering whether he had to be such a martinet. He had his own explanation: "Somebody has to be the bastard."

An Unexpected Detour

The enriching of uranium oxide is essential for the fueling of nuclear naval vessels and commercial reactors. Uranium oxide when mined contains 0.7 percent of the isotope U-235. Enrichment involves raising the U-235 content to three or four percent. U.S. uranium processing was conducted in three large plants: Oak Ridge, Tennessee; Portsmouth, Ohio; and Paducah, Kentucky. All three plants used gaseous-diffusion technology. Since World War II centrifuge technology requiring considerably less power had been developed and refined. Commercial ventures using this technology were under active consideration by the Ford administration. I was enthusiastic about the technology itself but not about its commercial application. I felt that it would be too difficult to maintain adequate safeguards while preventing nonproliferation within a private venture.

Knowing that this matter was to be discussed at Cabinet level, I called the White House to see if such a meeting was going to be held over Memorial Day weekend. We had planned to sail *Viva* from Chesapeake Bay to Manchester, Massachusetts, over the holiday. Once I received a negative response from the White House, we headed for Galesville, Maryland, where *Viva* was berthed. We were well up the bay by 11 p.m., when we spotted a vessel rapidly closing on us from astern: Coast Guard looking for Dr. Seamans! They put me in touch with Bob Fri, who said I was expected at the White House at 9 a.m. for a meeting with the President. A car would be available at 4 a.m. at the dock in Chesapeake City, which would take me to a local airport for a flight to Washington. A thick fog had descended by the time I reached the airport, and the place was deserted.

The driver was willing to drive all the way, but the time was short. We traveled at high speed down the eastern shore, over the Bay Bridge, and into Washington, arriving at the White House with minutes to spare. As President Ford came into the Cabinet Room, he shook hands all around, commenting when he looked at me that perhaps he had diverted me from sailing. President Ford decided to turn over the enrichment of uranium oxide to the private sector, a decision that I had to defend and that Congress ultimately rejected.

A Soviet Tour

Energy comes in many forms and is used in many ways, so it is not surprising that the United States has many foreign energy associations. The International Atomic Energy Agency (IAEA), headquartered in Vienna, was an arm of the United Nations that sought to minimize the proliferation of fissionable material. Another organization, the International Energy Agency (IEA), sought, by pooling the technology and resources of its member nations (oil consumers), to lessen the power of OPEC (Organization of Petroleum Exporting Countries) to control the price and volume of exported oil.

There were agreements bilateral and multilateral. One of these, which ERDA inherited, was the U.S.-U.S.S.R. Commission on the

Peaceful Uses of Nuclear Energy. Commissar Petrosyant and I were the cochairmen. During ERDA's first year the Soviets came here; during the second we traveled to the Soviet Union. The host country took care of all expenses. Prior to a trip, negotiations were extensive. The Soviets wanted to visit a number of classified laboratories and spend a weekend at Las Vegas. These requests were turned down. I got sick of their Las Vegas needling, but felt that was better than a congressional hearing on excess entertainment expenditures.

The following year our trip included Moscow, Shevchenko, and Yerevan at the foot of Mt. Ararat in Armenia. We went there to visit and discuss earthquake safeguards for two nuclear power plants. There was also a research center with a small nuclear reactor. Yerevan happened to be Petrosyant's home and the seat of the Armenian church. The Christofolis, the head of the Armenian church, was, according to Petrosyant, "a truly religious person." Our joint commission paid him a visit, which required two sets of translators since the Christofolis spoke only Armenian. The American and Soviet delegations both took the pledge that we were working together for the good of mankind.

During our business meetings in Yerevan, Gene had an even greater ecclesiastical experience. She was taken by the wife of the chief Soviet scientist for a day in the country. They had a driver and an interpreter and were accompanied by Petrosyant's daughter, who had been instructed to "look after" Gene. The Russian women suggested going to a fifteenth-century church, Gene being a "Christian person." So off they drove, up into the foothills and along dusty roads.

They finally came to a place with numerous streamers blowing in the wind, placed there by thankful believers. In front of them, they saw the church, which was mostly carved out of the hillside. They went into the vaulted cave. In the dim light the benches were lined up facing a very simple altar, where a bent, dusty priest was sweeping. It seemed appropriate to Gene to say a prayer, so she knelt down for a few moments and then approached the holy man. He had produced a few postcards and seemed very pleased to sell them.

As they entered the courtyard there was an excited crowd surrounding a splendid ram. The animal was decked with ribbons, and its horns were brightly painted. The ram was to be sacrificed by a grateful family who had recently been blessed with a son. The interpreter suggested staying for the ceremony "because Mrs. Seamans is a Christian person." The hostess did not agree and hurried the party back to the car.

All the way home, Gene was questioned. "How many churches are there in Washington?" "What animals do they sacrifice?" "What job do you have in the party?" "How many people live in your house?" "How many bathrooms do you have?" "Why do you change presidents?" And many more. Everyone in the car was eager for this chance to learn about us and our country. At the end of the day, Gene loaned the interpreter a copy of a *National Geographic* magazine that had an article on the U.S.S.R. She also gave him a small package of peanuts saved from an airplane. Next day he said, "My wife and I stayed up all night reading your magazine and eating the delicious nuts, one by one." He had never seen a foreign publication or tasted a peanut.

The Shah of Iran

Another interesting trip was to Iran, where I met with the Shah. He wanted nuclear power plants for his country and especially a uranium reprocessing plant. Our government, very concerned about the proliferation of nuclear weapons, didn't want him reprocessing uranium for weapons use. I was asked by President Ford to meet with him about this matter. Dick Helms, former head of the Central Intelligence Agency, was our ambassador to Iran at the time, and he was most troubled about the proposed meeting. He thought I was going to cause an international incident by trying to tell the Shah what he could or couldn't do.

The game plan was to try to get the Shah to take off his Iranian hat and don a regional hat and, specifically, to have him agree to have Iranian reprocessing done under the auspices of the International Atomic Energy Agency. With Dick Helms I finally met the Shah and gave him a letter signed signed by President Ford. He read the letter and said, "But of course this is what I want to do! Why would I want to have nuclear weapons? We're just a little country here. We're not going to fight the Soviet Union!"

On the way back to our embassy following the meeting, Dick said, "Well, at least he didn't throw you out of the office."

The Shah continued to press for nuclear reprocessing plants, of course.

Parading out of ERDA

On my last full day at the Energy Research and Development Agency, after two years of service and on the day before Carter's inauguration, I went down into the garage in the basement of our building and found four unusual cars. The only one I recognized, because I had driven it before, was a Chrysler with a turbojet engine. There were a couple of electric cars and a hydrogen-powered vehicle, as well.

I asked the attendant what the cars were doing there. "They're going to be driven in tomorrow's inaugural parade."

"Who's going to drive them?"

"The regular chauffeurs, I suppose."

"Well," I said, "I think I'll drive one of them."

To get into the parade was a complicated matter. There were three big rallying points—the marchers in one area, the horse-drawn vehicles in a second, and the floats in the third. Gene (who drove with me) and I were looked on as drivers of a float, so we went out in a very clear, cold dawn to the old naval station at Anacostia, where the float participants in the parade were mustered. There we spotted a number of celebrities, including Colonel Sanders (the Kentucky Fried Chicken man) wrapped in many blankets by his solicitous entourage and Kenny Stabler (the Oakland Raiders' quarterback who had just won the Super Bowl with his passing). We were bused from Anacostia to a cross-street near the Capitol, where our ERDA cars were waiting for us. The atmosphere was very festive. We were surrounded by the teams for different floats, including some gymnasts who grew tired of waiting and went through their whole routine right there on the street, probably to keep warm and nimble. When the parade finally got rolling, I drove the lead ERDA car, the turbocharged Chrysler, a powerful machine that made me feel as though I were taxiing out in an F-111. The pace of the parade was very slow, as it took President and Mrs. Carter a long time to walk down Pennsylvania Avenue, but finally we began moving into position. Some dogs were brought out to sniff under our vehicles for explosives. Then we were under way, heading down the avenue, our jet-age car directly behind a horse-drawn vehicle from the 1890s.

As we passed the reviewing stands, Gene would lean out one

window and I would lean out the other, both waving. At the end of the route we passed the presidential reviewing stand on Lafayette Square in front of the White House. Gene leaned across me, and both of us waved to President Carter and Vice President Mondale. Three blocks further on, we arrived at our route's end. We stepped out of the car, handed the keys to an attendant, and thereby ended my career at ERDA. I was a private citizen again and for good.

Dean at MIT

As soon as Jimmy Carter became President, his "energy czar," Jim Schlesinger, retained me as a consultant. I was allowed to remain in my office, which I occupied for about three months, working on a number of issues with my successor, Bob Fri.

During this period I got a call from Jerry Wiesner [former science advisor to President Kennedy], who had become president of MIT. He said, "Alfred Keil is going to be stepping down as dean of the engineering school. We have to go through various formalities here and I'm not offering you the job, but is this the kind of thing that might interest you?"

I said, "I doubt it, Jerry. I'm sure it involves a lot of administrative work, and that's the last thing I want to be involved in for a while. I've had it."

"Well, come on up and talk about it anyway." So I traveled to Cambridge and talked with Jerry and the provost, Walter Rosenblith. I also talked with Al Keil and realized my fears were well founded. Being dean meant looking at a thousand nitty-gritty details—not what I wanted at the time. When I made it clear to Jerry, Walter, and Howard Johnson (now MIT chairman) that I wouldn't accept the post of dean, I was offered instead an institute chair, one that is not tied down to any single department. There are eleven such chairs at MIT, but the Luce chair is unique. Its full name is the Henry Luce Chair for the Environment and Public Policy. Nobody had held it for a year or so. The idea had appeal, but was I really an environmentalist? I had had to wrestle, particularly at ERDA, with a lot of environmental issues, but in the eyes of the environmental movement I certainly would not have been considered a key player. Nevertheless, after consulting with Gene, who was as ready as I was to leave Washington, I finally accepted.

We sold our house in Washington, though it took some time. Again it was wrenching to leave this pleasant home on a dead-end street in Cleveland Park. We had a dinner-dance-cum-going-away party at the Chevy Chase Club with about 150 friends, including associates from NASA, the Defense Department, and the National Academy of Engineering. Then we left town, making a good clean break. I still see people today, however, who ask, "Do you and Gene still have your house in Washington?"

We were delighted to settle back in Cambridge. I started digging into what I might accomplish with my faculty appointment at the institute. Carroll Wilson asked me to join forces in a course on national and world energy issues. Carroll had been the first general manager of the Atomic Energy Commission. I also got involved with his World Coal Study (WOCOL), a big project on world coal supplies and usage.

I had also agreed to help Jerry Wiesner and Howard Johnson find a dean for the engineering school. Howard called me early in 1978 and invited me over to the Tavern Club for lunch. We chatted about this and that. "Well, now that you've got Washington well behind you," he said, "how do you feel about administrative work? Do you miss some of the things you were doing?" I had been feeling that I might be ready to get into the nitty-gritty again, and I told him so.

About three days later I got a call from him, asking me to come over to his office to chat with Jerry Wiesner and himself. I suspected that they wanted to ask me again to be dean of the engineering school, our efforts to find someone having so far failed. There were eight departments and four laboratories affiliated with the engineering school, making twelve heads reporting to the dean. These twelve were all more or less of the same vintage. To pick any one of them to be boss of the others would have been difficult. To bring somebody in from the outside also would have been tough, since Jerry was going to be stepping down within two years, and it would have been hard for an outside person not to know to whom he or she would be reporting in the near term.

The long and short of it was that I agreed, but in my formal letter of acceptance I asked for the option of stepping out at the end of two years. When Paul Gray succeeded Jerry after two years, I told Paul I would stay on for another year. It was a wonderful opportunity for me to get to know MIT better after being away from it, except for a brief stint in 1968, for more than two decades. There were still people who knew my name when I passed them in the corridor, but there had been tremendous changes in my absence.

One of the responsibilities of a dean is to bring faculty appointments to the academic council, a body unique to MIT. It includes the president and the deans, the provost, and a couple of associate provosts. It meets every Wednesday for the morning and through lunch, grappling with all the major issues facing the institute. A lot of its time is spent on faculty appointments—the straightforward way to change the face of MIT. Before going to the academic council, an appointment had to be cleared first by the individual's department, then by the school. The engineering-school council is made up of twelve department and laboratory heads and is chaired by the dean.

One of the cases I was involved with was particularly interesting, a case involving tenure for a woman in ocean engineering named Judith T. Kildow. Ocean engineering was once known as naval architecture, but was expanded to include platforms at sea, some coastline problems, and other newer matters. Judy had been working in the policy area, not in engineering per se. Hers was thought to be a marginal case when it came before the engineering council. It was finally supported there, though most of us thought it would not pass the academic council.

One strike against Judy was that she had changed jobs quite a few times. I looked into this and realized that she had done so in order to follow her husband, whenever he had relocated. A second strike was what seemed to be gaps in her employment record, times unaccounted for. When I checked I found that the gaps coincided with the births of her children. I came to believe that Judy deserved tenure and should not be disqualified because of family matters. Getting her tenure approved by the academic council was one of the really satisfying things that happened to me while I was dean of the engineering school. Judy is still teaching ocean engineering there and doing a great job.

Being dean of the engineering school brought my whole career into focus. I realized that my expertise was not in any technical area—the design of aircraft control systems, for example—but rather in managing technological organizations. Such management poses special problems of its own: for an inventor trying to bring new ideas to market; for a small company that needs large financing; for research laboratories that are not closely integrated with product divisions; for government trying to acquire high-tech systems within cost estimates; and so on. In the final analysis, an innovation can bear fruit only when the application is technically and financially feasible.

I couldn't see any place at MIT where these matters were being studied, except in the Sloan School where the discussion had little engineering or technical content. So I started working with Bill Pound, the Sloan School dean, to inaugurate a one-year master's program in the management of technology. As I got into it, I found that in almost every department in the engineering school, there was at least one person dealing with entrepreneurial or policy-type issues, Judy Kildow being one of them. The question was, could we get them all interested and participating in this new program? We finally pulled it together, and in my last year as dean, the first fifteen people came to MIT to take the program.

In the fall of 1992 I was sitting in my office when a student named Amanda Chou came in. She said she was interested in writing her master's thesis on the possibility of commercializing space activity. Specifically, she was considering the development and production of rockets by commercial entities and the selling of these rockets in a competitive market. I asked her what program she was in.

"The management of technology," she answered.

"How's the program going? How many are in it this year?"

"There are about fifty-five or sixty people taking it each year now," she answered. It was gratifying for me to find that the program was still alive and prospering.

I stayed on the MIT faculty as Henry Luce professor for another year after resigning as dean. I had never taken a sabbatical in my on-again, off-again MIT career, so I took one during the second half of that year, 1984. I used the sabbatical to begin the process of examining, filing, and in some cases dispersing the material and intellectual property I had accumulated over the years.

Jack Kerrebrock, head of the "aero and astro" department, said to me, "I don't like the idea of your leaving the institute entirely, Bob. You ought to have an office here, at least." He told me about the possibility of my being made a "senior lecturer," a position reserved for a small number of professors emeriti. I had been talking with my brother Peter (who was thinking of retiring from his law firm, Peabody Arnold) about taking an office together. Just about that time, he and his partners decided to move from their offices at 1 Beacon Street to new offices at Rowes Wharf, so Peter decided he would stay with the firm until mandatory retirement. I therefore decided to keep my office at MIT and soon accepted an appointment as senior lecturer. I retired from this position in June 1996.

One area in which I did substantive work in this capacity was to help organize Technology Day, the institute's version of homecoming. Technology Day is held during the week after graduation and is attended by large numbers of alumni. It includes the usual alumni functions—outings to the Boston Pops, dinners, dances, and so on. But unlike homecoming at other schools, Technology Day has a different, specific agenda each year.

I first got into it in 1989, when Paul Gray asked me if I would head up the Technology Day effort on the twentieth anniversary of the Apollo 11 moon landing. I recognized the large role MIT had played in the Apollo effort, and I thought it would be valuable to collect memorabilia and organize lectures examining this role. I thought we ought to invite as the keynote speaker an astronaut who had a degree from MIT, as more astronauts have been graduates of MIT than of any other institution except the U.S. Naval Academy. We finally chose Captain Frederick Hauck, commander of the first space shuttle flight after the *Challenger* disaster. The program started with an abridged NASA documentary on the lunar landing. Then, following Captain Hauck, faculty from several departments discussed the relevance of their research to space exploration and manned spaceflight. The program was a sellout.

Venture Capital

In my first year back at MIT (1977–78), Jerry Wiesner called me to say that Governor Michael Dukakis had started a commission to see if the state of Massachusetts couldn't be more supportive of new commercial enterprises. Jerry said he had been on it, but that he was now too busy at MIT. Was I willing to take his place? When I looked into it, I found it fascinating because it concerned my area of expertise—the commercialization and management of technology. Just about the time I said I would do it, Dukakis had the commission converted into the Massachusetts Technology Development Corporation (MTDC). It was authorized to provide venture capital in the form of loans or equity to companies in the state of Massachusetts. One requirement was that the start-up be unable to get funds any other way. Another was that it be located in a disadvantaged area where the new outfit would create employment opportunities.

I became a member of the MTDC board, and when Edward King replaced Dukakis, I was asked to stay on. I remained there until well after Dukakis's comeback in 1982. Two professional analysts selected about one in fifty applications for the attention of the board. Several of us would then go and meet with the individuals involved to see what they were doing. Once we had agreed to put some funds into a start-up, the company almost invariably got additional funding from other sources. We had a revolving fund of about three million federal dollars, which we could use to make loans or equity investments. If the investment worked out and we made a profit, the proceeds went back into the revolving fund, as did any interest on the loans.

While I was there, we helped on the order of thirty-five companies get started. If I'm not mistaken, thirty of them are still going. Some of them, like Interleaf, a software company, have done extremely well. The MTDC is still in business and doing an effective job.

More recently, I've been on the advisory committee of a private venture capital operation, Morgan Holland. My two brothers and I made a modest investment in one of their funds. It had some similarity in my mind to the MTDC. I found that the entrepreneur with the hot idea had great difficulty staying with it once the company got to a certain level of activity. There was almost always a traumatic period of transition when an outside manager had to be brought in to run the show and when the biggest question was what to do with the original entrepreneur.

My brother Peter was the lawyer in three classic cases involving Henry Kloss, the brilliant inventor behind KLH, Advent, and Kloss Video. In all three cases, Peter found, Henry was able to bring products along and sell them—up to the limit of about $10 million in annual sales. Beyond that, he could not do it all himself; yet he would refuse to step back and let somebody else take over some of his responsibilities.

Johnny Appleseed's

Another commercial involvement was with a Bosson family business interest, the Boston and Lockport Block Company, originally suppliers of pulleys or "blocks" for America's sail-powered merchant fleet. It was first located in East Boston, by virtue of the fact that Gordon McKay, famous designer of clipper ships, had his vessels built there. It merged with the Lockport Block Company and eventually designed blocks for other uses, including freighters and drilling rigs.

The company had its ups and downs. As a result of a near-bankruptcy and my Grandfather Bosson's presidency of the County Savings Bank, my mother's family ended with approximately a third interest in the company. Uncle Campbell Bosson took over management of the family interest when Grandpa died. In time, my brother Peter took over from him, and my brother Donny became president and general manager.

The company had not done a very good job of reinvesting its profits in new machinery and technology, and Peter and Donny together tried to revitalize the company—against the wishes of the other two families who had significant interests. In time, my brothers managed to wrest control of the company from these families by buying out their shares. My brothers were fortunate to have Frank Browning, a very astute person, as an outside director. Frank pointed out that the company had a large amount of cash sitting on the sidelines and suggested that the company put it to work. Perhaps the natural solution would have been to buy a cable or rope company to complement the block operations, but Frank Browning steered the effort in an entirely different direction.

He found that Sam Batchelder, our neighbor and my sailing partner, might want to divest ownership of Johnny Appleseed's, the clothing retailer, while retaining the management. This proved to be the case, and Peter and Donny bought the company at a most opportune time. The block business became nearly moribund with the advent of the container ship, but Peter and Donny managed to sell the block business to a Chicago firm before the bottom fell out. This gave them additional funds to invest in Appleseed's.

Sam Batchelder stayed on as president at Appleseed's, a role in which he exercised his real expertise, buying. He had a tremendous flair for knowing what people wanted. Jim Shaughnessy, an excellent financial officer who had worked with Donny at Boston and Lockport, moved into the Appleseed's operation and did a fine job reconfiguring the financial end of the business. Frank Browning died, and his son, Frank Jr., successfully replaced him. The company flourished under its new ownership and management. It had entered early into the mail-order business and did extremely well there. It also successfully expanded the number of stores and outlets.

There came a time when a decision had to be made. Other upscale retailers like the Talbots were aggressively expanding and reinvesting large sums in their operations. It became clear to us that to stay in the ballgame, we would have to make another round of major investments. We were all getting older—Peter, Donny, Frank Jr., and I—and didn't feel up to the task of raising the necessary capital and pushing the operation to a higher level. We decided the company could be much better served by new ownership, and we let it be known in the business community that this was our intention. We received inquiries but felt the company was worth much more than we were being offered.

One wonderful day my brother was in his office when he got a call from a New York business broker who had been working with him on the sale. "I've got someone from Switzerland in my office here in New York who wants to come up and chat with you this week about the possibility of buying Appleseed's."

"I'm so busy," Peter said. "I don't think that would be possible."

"I think it would be worth your while," the broker said.

The man visited Peter the next day, and they spent a couple of hours together. The man then asked to see the company, with the proviso that he speak with Peter again when he was through with his visit. While the man was off seeing the company, Peter did some back-of-the-envelope calculations and decided that a high but fair price for the company would be about twice what he had been offered previously. When the man from Switzerland returned, he came up with the same number! Within a couple of months, we had an offer in writing with one condition attached. The deal was dependent on growth continuing at the same pace for another quarter. We watched the numbers (very closely!) for the next three months, and fortunately they lived up to all expectations. The Swiss company bought us out.

In retrospect, we got out of the mail-order business at the peak of that market. With our successful exit from the block business just before that collapsed, we seem to have been one step ahead of the sheriff all the way along.

Directorships

In 1968, just after leaving NASA, I joined the board of a corporation for the first time, when I became a director of Aerospace Corporation. A Los Angeles outfit, Aerospace came into being as a spinoff of Ramo-Woolridge, itself a predecessor of TRW, Inc. Ivan Getting, formerly of MIT and Raytheon, was its president, and the board was a remarkable collection of people, including Dr. James H. ("Jimmy") Doolittle, Dr. Charles C. Lauritsen, and Frederick R. Kappel, the chairman of AT&T. I left the Aerospace board as soon as I joined the Air Force, then returned in 1977 upon leaving government service. In 1981, I became chairman, retiring three years later.

Other boards I've sat on include Pneumo Company, which made pneumatic actuators used in aircraft; the Lord Company in Erie, Pennsylvania, which makes shock mounts; and Air Products Corporation, which sells liquid oxygen, liquid hydrogen (fuel for the Saturn rocket and the space shuttle), and other specialty liquids and gases derived from air. Keith Glennan, also on the Air Products board, got me involved while I was at the National Academy of Engineering. While at the Energy Research and Development Agency, I was not allowed outside interests, so I left Air Products, only to return as soon as I had left ERDA. Air Products came from a single family, the Pooles of Allentown, Pennsylvania. It was a very well-run scientific-based organization, a perfect example of how to keep growing rapidly without going heavily into debt. They reinvested a large percentage of profits in research and development (R&D) in an intelligent fashion that resulted in new and improved products.

Eli Lilly and Company

A much larger, better-known example of this phenomenon is Eli Lilly and Company, a family-run business that, through R&D, became one of the world's largest pharmaceutical firms. Yet by 1975 Lilly, a multibillion-dollar company, had no outside directors. Company lawyers finally told Eli Lilly, the old gentleman himself, that he had to have outside directors. Why? He knew how to run the company! The lawyers reminded him that his was a public company, and that in order to be responsible to all share-holders, the board needed outside representation.

Lilly finally agreed on the condition that the outsider be William McChesney Martin, just-retired chairman of the Federal Reserve Board. A brilliant man (and superior tennis player), Martin had been head of the New York Stock Exchange at age thirty-two. But Bill Martin said he couldn't go along with being the only outside director. Lilly had to have at least one more, so the company picked as its second outside director Albert L. Williams, the chief operating officer of IBM. When George Bush temporarily left the government during the Carter administration, he became the third.

When Bill Martin reached seventy, mandatory retirement age, a replacement had to be found. Bill, who had known me through the National Geographic Society, suggested me. Richard Wood, Lilly's chief executive officer, gave me a call, saying he'd like to chat with me in Washington. When we met, I said, "I don't know anything about phar-maceuticals. What can I do to help you?"

He said that he felt my government experience and my expertise in managing technical endeavors would be of benefit to the company. Eli Lilly proved to be a great firm to work for. It was definitely an ingrown company, but it had wonderful teamwork. Things went along swim-mingly until the early 1980s, when the company brought out a new prod-uct called Oraflex, an anti-arthritic drug. It was a nonsteroid that seemed to have remarkable characteristics. People who hadn't played golf for years were taking it and suddenly shooting in the 90s! Oraflex was the first drug Lilly had marketed directly to the public rather than through doctors. The company expected that it would be one of its biggest mon-eymakers ever. A big one to Lilly was a product that grossed over $100 million a year. Oraflex looked as though it might do two to three times that if properly marketed. The company hired a Madison Avenue agency and placed ads in print and on television.

The drug had been brought to market in Great Britain about a year before getting Food and Drug Administration (FDA) approval in this country. Within a month of the FDA's approval, a Scottish doctor wrote a letter claiming that patients for whom he had prescribed Oraflex had died with kidney problems. The British agency that regulated pharmaceuticals immediately took Oraflex off the market. It soon was taken off the market in this country as well, but not before enough people had used the drug to trigger more than $100 million in lawsuits. It was a bad situation, the kind of thing that turns up on *60 Minutes,* with a woman in Scotland crying at her mother's gravesite. A grand jury was ordered to investigate.

Lilly's Washington law firm suggested that the company empower its outside directors to investigate what had happened and to fire any member of the corporation deemed responsible for improprieties. By then there were six outside directors, and I was made chairman. We got Cyrus Vance as our legal counsel and hired outside medical experts as well. A junior assistant to Vance interviewed people in this country and in England to find out what really had happened. The company had known from tests that there was a lesser problem with the drug. Between 5 and 10 percent of all users became very sensitive to the Sun and experienced a violent rash. The label had printed warnings to this effect. But the alleged renal complications took the company completely by surprise.

We found that all of the Scottish patients who had allegedly suffered from taking Oraflex had been taking a number of other drugs in addition, so that it could not be proved that Oraflex had been the agent of death. Still, there was some question as to whether Lilly's director of research in England had been forthcoming about his findings prior to FDA approval, even though this was not a legal requirement. The result of our deliberations was a report delivered by me (with Cy Vance at my side) to the full board in Indianapolis. When we had finished, Dick Woods said, "That's the last time Lilly will market a drug publicly. We'll never do it again." The director of research in England was fired. Fortunately for the company, the grand jury investigation was called off—in large part because the company was able to show that it had conducted an in-house investigation in good faith. The lawsuits were settled one by one.

Combustion Engineering

Combustion Engineering was a company I learned something about while at the Energy Research and Development Agency, because at the time it was one of four companies in the nuclear business—General Electric, Westinghouse, and Babcock & Wilcox being the other three. Arthur Santry, who grew up sailing in Marblehead, practiced law with his father in Boston for a number of years until his uncle, the chief executive officer of Combustion Engineering, asked him to join the company as in-house counsel. Eventually, the directors felt it was time for the uncle to retire, and they recommended that Arthur replace him. The uncle was reluctant to leave, but ultimately his nephew was asked to assume the mantle.

Arthur got in touch with me just after I had left ERDA and asked if I would both go on the board and serve as a consultant. This would make me an in-house director. At the same time I was asked to go on the board of General Electric. I clearly couldn't do both because of the competition between the companies in the nuclear area. GE couldn't believe it when I turned them down in favor of Arthur's company. I also had to quit the Air Products board when I joined Combustion because Air Products had acquired a design and construction company in Philadelphia that serviced nuclear plants.

At Combustion I felt I had a responsibility to do something other than go to board meetings. I started looking into the company's research and development effort and was disappointed to find that it was rather minimal—other than in the nuclear area, where there was competence and an interesting research effort. Combustion spent less than 1 percent of sales on R&D. Charlie Leaper, with a doctorate from MIT, emerged as the most respected technical leader at Combustion and was made vice president for engineering. Charlie coordinated the company's R&D effort and was given a small budget to enhance research in the various divisions. A board committee, which I chaired, met with Charlie and the divisions on a rotating basis to review research results.

Arthur did a remarkably good job with the company. It grew while he was CEO, especially after he got the company into the oil and gas services business in time to catch the oil boom of the late 1970s. There came a time, however, when the board started getting restive about Arthur's successor. Arthur was getting on in years, and though he did have an excellent chief operating officer, James Calvert, a retired admiral who had skippered the first nuclear submarine under the arctic ice pack, the board didn't consider him CEO timber because he lacked financial experience. Arthur finally got serious about finding an outsider and picked the number-three person at AT&T, Charles Hugel. We thought we were lucky to get him.

Charlie Hugel was an entrepreneur with a strong strategic sense. He recognized the limitations of Combustion's nuclear, oil, and gas businesses and bought Taylor Instruments and several other smaller businesses in order to generate a new base of operations. However, I saw my efforts to consolidate the company's research effort go up in smoke when Charlie Leaper was terminated. Without warning, Hugel also fired a number of key people on Arthur's staff, including the very capable Jim Calvert. Then Hugel was asked to become chairman of RJR Nabisco at the time that company's financial shenanigans were coming to a head. This limited the time he had available for Combustion Engineering.

Combustion Engineering's profits became losses. When Asea Brown Boverei (ABB), a Swiss-Swedish company, appeared over the horizon with a buyout offer, the directors listened carefully. In a matter of two months we sold out to them-a good move, to my mind. Charlie Hugel became vice chairman of the ABB board, but didn't stay there long. All in all, though, it was a sad ending for Combustion Engineering. In defense of Charlie Hugel, the business was rapidly changing, becoming more global and competitive. Strong medicine was needed.

Putnam Funds

While I was dean of the engineering school at MIT, Bill Pounds called me one day. I figured he wanted to talk about the management-of-technology program we were collaborating on or about something else at MIT. Instead, he began talking about the Putnam Funds, of which he was a director and the chairman of the nominating committee. He said his committee hoped that I would be willing to go on the board.

"I really don't know anything about the investment business

except through my own personal participation in it," I said. "The sub-
tleties of it are entirely foreign to me."

He suggested that I chat with George Putnam. George pointed
out, when we met, that technology was having such an impact on the
development of new companies that it was very helpful to have someone like me on the board. Vannevar Bush, former dean of the engineering school at MIT, had filled this role for Putnam. During World
War II, Bush, as head of the Office of Scientific Research and
Development, had reported directly to President Roosevelt on all science and engineering for the war effort from his office at the Carnegie
Institution of Washington. After the war he wrote a book, *Science, the
Endless Frontier,* that generated a tremendous amount of favorable
publicity and led to the establishment of the National Science
Foundation. In the world of technology, he was top of the line. George
wanted me to replace him.

"That," I told George, "is like saying you've had Ted Williams on
your team and you're going to replace him with the Red Sox batboy!"
But I appreciated the compliment and did agree to serve with Putnam.
It was a great education. I joined just when mutual funds were starting
to grow. When I arrived Putnam had fewer than ten funds with $5 bil-
lion in assets. When I left it had seventy funds with $45 billion in assets.

The board had about a dozen trustees who met once a month on
Thursdays at two o'clock. The meetings ran through dinner, then
resumed the following morning, running through lunch. One of the full
board's many responsibilities was voting dividends for each of the funds.
Between board meetings there were committee meetings. One committee
was responsible for negotiating annual management fee with Marsh
McClennan, which marketed the Putnam Funds and managed its operations. There was also a legal committee and a proxy committee, which I
chaired. It was our job to determine a rationale for voting the proxies
received each year by the Putnam Funds as shareholders in many thousands of companies. All in all, there was a tremendous amount of detailed
work involved in all this. I complained periodically that there had to be
a more efficient way to operate the board. Finally, I was made chairman
of a small governance committee. I found that very interesting, and our
committee came in with a series of recommendations to make the board
more efficient. In this way, I felt I left my small mark on Putnam.

As chairman of the proxy committee, I also became involved in a number of interesting situations. In the 1985–87 timeframe, when leveraged buyouts were happening almost daily, shareholders were being asked to vote for or against them. Often a fund manager with a large stake in a company would favor the takeover on the grounds that current company management was not doing a good job. Of course, the fund manager liked to see the stock price shoot up, as a result of a takeover. Standing back a bit, one could say that in many cases a buyout was not good for a company-that many were being effected simply to break up companies, sell off their assets, and fire a lot of employees. In the long run, the value of a takeover target might actually decline, and this would ultimately be reflected in the stock price.

Bearing all this in mind, were we going to support the management of Gillette, one of the biggest companies in Boston? A group of financiers in the Henry Kravis [a corporate takeover specialist] mold began negotiating to take over the company. I felt that if they succeeded, Boston wasn't going to end up with anything more than a couple of old buildings with broken windows. The Putnam fund manager with the largest stake in Gillette was all for the takeover. As chairman of the proxy committee, I went against him (after consulting with George Putnam), and we decided to vote against the takeover. The final shareholder vote was extremely close, so that the voting of our large block of shares in support of management had an important bearing on the outcome. Soon afterwards, Gillette came out with the Sensor razor, and the company's growth accelerated.

Nonprofits

During the last twenty-five years, especially since my return to MIT, I have had an inside view of a number of nonprofit organizations by serving as a director. My first major experience with a nonprofit was with the National Geographic Society in Washington. Founded in 1888 by Gardiner H. Hubbard, Alexander Graham Bell's father-in-law, the *National Geographic* had been passed on in the Grosvenor family, descendants of Bell's daughter. At the time I became associated with it, the flagship magazine was being edited by Gil Grosvenor, Bell's great-grandson. His father, Mel Grosvenor, a wonderfully gregarious, enthusiastic person, was chairman emeritus, and was still pulling down a salary. Mel Payne, who had been brought in by the Grosvenors, was chairman. The society had grown to the point where it could no longer be looked on merely as a family preserve, so a committee on governance was formed to analyze the succession question. The committee included Conrad Wirth, former director of the Park Service; former chairman of the Federal Reserve Board William McChesney Martin; Jim Webb; and a couple of others, including myself.

We determined that management of the society needed to be broadened out beyond the Grosvenor family. Not only did this make common sense, but also the *Geographic* was getting involved in areas other than its publications. It was producing television programs, organizing tours, distributing *National Geographic* merchandise, and so on. We recommended that instead of Mel Payne, an outsider (perhaps a sitting board member) be made chairman. We further felt that Gil Grosvenor, if promoted to president, ought not to continue as editor of the magazine. Finally, concluding that Mel Grosvenor's salary might appear unseemly, we recommended that it be phased out over a three-year period.

The board meeting at which our recommendations were discussed was one of the most dramatic I've ever attended. Conrad Wirth, chairman of the committee, wasn't there that day, so there was no single spokesman to explain the committee's decisions. We probably hadn't done enough to prepare everybody for our recommendations. Suddenly we were the devils. Mel Payne said he would never again put together a management committee, claiming that we were trying to sabotage the *Geographic*—taking a great institution and throwing it to the winds.

In the end, we won and we lost. Mel Grosvenor agreed to retire, and Gil agreed to take over the presidency without the editorship. Mel Payne, however, remained as chairman. Ultimately, Gil became chairman and president. He reviewed the society's strengths and weaknesses, while taking action to strengthen the society and to select goals appropriate to the twenty-first century.

One of the perquisites offered by the *National Geographic* has been the chance to participate in trips to remote places, staged by the research committee. The National Geographic Society gives grants to worthy projects around the world, and every three or four years members of the research committee visit some of these work sites. Gene and

I have seen the dawn at Machu Picchu, the historic sights of mainland China, the ruins of Jordan and Persepolis, a Bedouin camp in Israel, the mosques of Iran, the island of Santorini, the wildflowers of Crete, the skeletons of Herculaneum, the kangaroos of Australia, the glaciers of New Zealand, the strange birds and marine life of the Great Barrier Reef, and many other remarkable sights. Sometimes with us on these trips was Lady Bird Johnson, with her security entourage. We have felt really fortunate to be a part of this lively, knowledgeable group.

In addition to the *National Geographic,* I have been involved with the National Cathedral in Washington (chapter member), the Carnegie Institution of Washington (trustee and vice chairman), the Boston Museum of Science (trustee), the Woods Hole Oceanographic Institution (trustee), the New England Medical Center (board of governors), and the Trustees of Reservations (standing committee). I have thoroughly enjoyed my association with each, particularly knowing the participants and especially the movers and shakers. All of these organizations have had management challenges, and all have held major fund drives.

Finally, let me mention the Sea Education Association (SEA), which has classrooms and laboratories aboard two 130-foot schooners. Jimmy Madden got me involved soon after I returned from Washington in 1978. I was asked to serve first as a trustee and then as chairman (1989–1993). SEA was originally run out of a church basement in Woods Hole. In time, it acquired a ten-acre campus and built some houses around a small central building—enough to accommodate fifty students at a time living about eight to a house. I thought we needed a proper teaching facility ashore so I went to Tom Watson of IBM, a great friend of Jimmy Madden, and asked if he would be interested in supporting our effort to establish a marine center in Jimmy's memory. He sent me one of the nicest letters I've ever received, offering a substantial matching grant to get our fund-raising effort moving. He spoke at the dedication of the center in June 1993.

Write a Book?

Gene and I have been married over fifty years, and our family now numbers twenty-four, ages one to seventy-seven (myself). Our children and grandchildren are scattered, but we all enjoy each other's company as often as possible. Traveling from Berkeley, Pittsburgh, Philadelphia, Cambridge, and Beverly Farms, most of us still meet annually for spring skiing at Vail. Everyone is welcome for Christmas and summer pleasures at Sea Meadow, where Gene and I live year-round. Fortunately, the house holds us all. Our offspring's combined energies, desires, and special interests make these visits memorable. I don't know what we'll do when great-grandchildren enter the picture!

Since 1960 I have never completely lost contact with NASA, the great adventure of my professional life. While I was at the Air Force, I served on the President's Space Council, chaired by Vice President Spiro Agnew and charged with determining what NASA ought to do as a fol-low-up to the Apollo program. As administrator of ERDA, I was closely involved with NASA through ERDA's research work. I tried very hard to get NASA involved in energy issues. My next involvement with NASA came after the *Challenger* accident, when I chaired the effort to set up a NASA Alumni League.

More recently I was chairman of a special White House Committee on the Space Station. I also served on a space station com-mittee chaired by MIT president Dr. Charles ("Chuck") Vest. Serving with me on the latter was Dr. Edward B. Fort, chancellor of North Carolina Agricultural and Technical State University in Greensboro. He and I were chatting one day, when he asked, "Did you really know any of the astronauts?"

"Yeah," I said, "I knew quite a few of them."

"Did you ever meet Wernher von Braun?"

"Oh, sure, I knew Wernher."

He asked me for supper that evening to inquire further into my days at NASA. By the time supper was over, I had recounted several of the stories in this book. Afterwards he said, "That was worth the whole trip to Washington. Have you ever thought of writing a book?"

"I've given it a little thought," I said.

Selected Documents

☆   ☆   ☆

As head of the executive branch of the federal government, the President of the United States has responsibility for a dozen major departments (the largest being the Department of Defense), as well as a number of independent agencies. Of the agencies, the National Aeronautics and Space Administration (NASA) and the Energy Research and Development Agency (ERDA) were prominent during my days in government. ERDA became part of the Department of Energy in President Carter's administration; NASA, although occasionally the center of criticism, continues to function on its own.

The President's responsibilities extend well beyond the departments and agencies to a wide variety of boards and commissions. Reporting to the President are on the order of 150 individuals, in all. Such a span of control in the management of a large enterprise is an order of magnitude larger than that dictated by the accepted wisdom. To obtain some degree of conformity throughout the executive branch, the President has an Executive Office employing about 1,000 personnel. Included within this inner circle is the Office of Management and Budget (OMB).

The President has the authority to hire (with the consent of the Senate) and fire about 1,000 key members of the administration. Within NASA, only two positions are filled by formal presidential appointment, those of the administrator and the deputy administrator. When I first joined NASA, I held an "excepted position," serving at the pleasure of the administrator. Subsequently, I received presidential appointments to three positions: NASA deputy administrator, secretary of the Air Force, and administrator of ERDA. The correspondence in this appendix relates to my appointments and my relationships with the Office of the President, as well as a number of key reports on policy and project issues.

July 15 and 26, 1960: NASA Administrator T. Keith Glennan's letter to me and my response

Following our dinner meeting on Monday, June 27, 1960, I received a midweek call from Keith in which he indicated that he needed an early decision and was having final conversations with another candidate. After several soul-searching discussions with Gene, I threw my hat in the ring by week's end. We had planned a summer cruise aboard Serene. *We sailed out of Manchester harbor with Bill and May English immediately after receiving the affirmative phone call referred to in Keith's letter but before any announcements had been made. By the time we reached the coast of Maine, the word had spread, and friends and acquaintances were offering congratulations. Upon my return, I received his letter with its somewhat ominous final paragraph. I sent him my acceptance on July 26, in which I mentioned a two-year stint for "family planning purposes." On this and all subsequent jobs, I have offered my services for two years, a period in which much can be accomplished but with a foreseeable end. Perhaps I was always a bit concerned about my loss of freedom.*

15 July 1960

Dr. Robert C. Seamans, Jr.
675 Hale Street
Beverly Farms, Massachusetts

Dear Bob:

This will confirm our telephone conversation of this morning. I am attaching for your files a copy of the press release which is to be made on Tuesday morning next. It is my understanding that this action has been coordinated with the action to be taken by the Public Relations Department of RCA in announcing your departure from your present job.

I was delighted to learn that you could spend some substantial portion of Wednesday, 27 July, in Washington with Dick Horner. As I told you, he will want to help you in connection with your office assistants to the greatest extent possible. It seemed best to set this kind of a meeting for some day other than those scheduled for the NASA-Industry Conference. In this connection, our Advisory Committee on Organization will be meeting on the 28th and 29th, and, as I told you, Dick Horner and I plan to take dinner with them on the evening of Thursday, 28 July. I would like very much to have you join us for that session which I think will be very interesting. It is my understanding that you can do this.

I think I should confirm in writing my offer to you of the post of Associate Administrator of NASA, reporting to me, at the annual salary of $21,000. In this post you will have responsibility for operating management of the Agency in the same manner as has Mr. Horner. I am delighted that you will join us and I look forward with eager anticipation to your being on the job full time. In the interim, as circumstances may permit, we will look forward to having you with us from time to time.

I am going to be away for ten days but Dr. Dryden will be here during my absence and can answer any questions which you might have. I hope that your holiday is giving you plenty of rest and building up a substantial store of energy—I can assure you you will need the latter.

Kindest personal regards.

Sincerely,

T. Keith Glennan
Administrator

July 26, 1960

Dr. T. Keith Glennan

Dear Keith:

Thank you for your letter of 15 July. I appreciate the opportunity to discuss office assistants with Dick Horner on Wednesday, 27 July, and I am happy to have dinner with you, Dick Horner and your Advisory Committee on Organization on Thursday evening, 28 July.

As I indicated in our telephone conversation of 9 July, I am looking forward to working with you as Associate Administrator of NASA, starting 1 September at an annual salary of $21,000. Although I recognize that many factors may affect the duration of the assignment, we are thinking in terms of a two-year period for family planning purposes.

With best regards,

Sincerely,

Robert C. Seamans, Jr.

☆ ☆ ☆

May 8, 1961: Letter, with report attached, from Defense Secretary Robert McNamara and NASA Administrator James E. Webb to Vice President Lyndon B. Johnson

During the last week in April and first week in May, the Vice President held a number of meetings relevant to his charge from President Kennedy. Meeting with him was a disparate group, ranging from those in charge of NASA, such as Jim Webb and Hugh Dryden, to Wernher von Braun, Air Force General Bernie Schiever, and others in and out of government. Finally, the Vice President called directly on McNamara and Webb for their specific recommendations. The important sections of the resulting, somewhat convoluted, document are included below. It should be recognized that the Department of Defense (DOD) already had proposed a draft report containing many ideas and recommendations not germane to the joint decisions. These had to be removed by negotiations, and joint NASA-DOD findings and recommendations had to be added.

The letter and report were delivered to the Vice President on the morning of May 8, just prior to the celebration for Mercury astronaut Alan Shepard. Following a White House award ceremony, Shepard delivered an address to Congress, overseen by the Vice President and House Speaker John McCormick. He then went with his family to the State Department, where he was tendered a luncheon by the Vice President. Johnson left the luncheon, report in hand, for a meeting with the President.

May 8, 1961

Dear Mr. Vice President:

Attached to this letter is a report entitled "Recommendations for Our National Space Program: Changes, Policies, Goals," dated May 8, 1961. This document represents our joint thinking. We recommend that, if you concur with its contents and recommendations, it be transmitted to the President for his information and as a basis for early adoption implementation of the revised and expanded objectives which it contains.

Sincerely,

Robert McNamara
James E. Webb

Introduction

It is the purpose of this report (1) to describe changes to our national space efforts requiring additional appropriations for FY 1962; (2) to outline the thinking of the Secretary of Defense and the Administrator of NASA concerning U.S. status, prospects, and policies for space; and (3) to depict the chief goals which in our opinion should become part of an Integrated National Space Plan. . . .

I. Recommendations for FY 1962 Add-ons

Our recommendations for additional FY 1962 NOA [New Obligation Authority] for our space efforts are listed below. They total $626 million, of which all but $77 million is for NASA. . . .

II. National Space Policy

Projects in space may be undertaken for any one of four principal reasons. They may be aimed at gaining scientific knowledge. Some, in the future, will be of commercial or chiefly civilian value.

Several current programs are of potential military value. Finally, some space projects may be undertaken chiefly for reasons of national prestige.

The U.S. is not behind in the first three categories. Scientifically and militarily we are ahead. . . .

III. Major National Space Goals

It is the purpose of this section to outline some of the principal goals, both long-range and short-range, toward which our national space efforts should, in our opinion, be directed. . . .

We recommend that our National Space Plan include the objective of manned lunar exploration before the end of this decade. It is our belief that manned exploration to the vicinity of and on the surface of the Moon represents a major area in which international competition for achievement in space will be conducted. The orbitiing of machines is not the same as the orbiting or landing of man. It is man, not merely machines, in space that capture the imagination of the world. . . .

The establishment of this major objective has many implications. It will cost a great deal of money. It will require large efforts for a long time. It requires parallel and supporting undertakings which are also costly and complex. Thus, for example, the RANGER and SUR-VEYOR projects and the technology associated with them must be undertaken and must succeed to provide the data, the techniques, and the experience without which manned lunar exploration cannot be undertaken.

The Soviets have announced lunar landing as a major objective of their program. They may have begun to plan for such an effort years ago. They may have undertaken important first steps which we have not begun.

It may be argued, therefore, that we undertake such an objective with several strikes against us. We cannot avoid announcing not only our general goals but many of our specific plans, and our successes and our failures along the way. Our cards are and will be face up—theirs are face down.

Despite these considerations we recommend proceeding toward this objective. We are uncertain of Soviet intentions, plans, or status. Their plans, whatever they may be, are not more certain of success than ours. Just as we accelerated our ICBM program we have accelerated and are passing the Soviets in important areas in space technology. If we set our sights on this difficult objective we may surpass them here as well. Accepting the goals gives us a chance. Finally, even if the Soviets get there first, as they may, and as some think they will, it is better for us to get there second than not at all. In any event, we will have mastered the technology. If we fail to accept this challenge it may be interpreted as a lack of national vigor and capacity to respond. . . .

☆ ☆ ☆

November 15 and December 6, 1961: Letter from NASA researcher John Houbolt to me, and my response

Even before President Kennedy's special message to Congress, NASA was switching into high gear. Many policy issues needed addressing. To what extent should NASA hire additional personnel? Where would we find individuals capable of managing large programs and projects? Where should the management of Apollo be located? Were new NASA centers required? To what extent could NASA enlist the support of the Army, Navy, and Air Force? Should the Apollo Saturn be assembled outdoors, as was customary, or indoors? How was NASA going to be managed to control schedules and costs or to select contractors?

None of the decisions were as intractable as the selection of the mission "mode"—the strategy adopted for getting humans and equipment to the Moon. There were two strong camps—one in favor of direct ascent, the other for Earth-orbit rendezvous. However, there was also a "voice in the wilderness," that of John Houbolt, who had a small team located at the Langley Research Center in Langley, Virginia. He advanced the concepts of lunar-orbit rendezvous.

Houbolt may not have felt confident that NASA was approaching the lunar landing "fairly and frankly," but my written response to him, given here, was a lot more supportive than my first reaction, that he should cease and desist. Houbolt may have been a "crank," but I thought his views made sense, and I kept checking to be certain Brainerd Holmes and his systems analyst Joe Shea gave Houbolt's views careful consideration.

Dear Dr. Seamans:

. . . Since we have had only occasional and limited contact, and because you therefore probably do not know me well, it is conceivable that after reading this you may feel that you are dealing with a crank. Do not be afraid of this. The thoughts expressed here may not be stated in as diplomatic a fashion as they might be, or as I would normally try to do, but this is by choice and at the moment is not important. The important point is that you hear the ideas directly, not after they have filtered through a score or more of other people, with the attendant risk that they may not even reach you.

Manned Lundar Landing Through Use of Lunar Orbit Rendezvous

The Plan.—The first attachment outlines in brief the plan by which we may accomplish a manned lunar landing through use of a lunar rendezvous, and shows a number of schemes for doing this by means of using C-3, its equivalent, or even something else. The basic ideas of the plan were presented before various NASA people well over a year ago, and were since repeated at numerous interlaboratory meetings. . . .

Regrettably, there was little interest shown in the idea—indeed, if any, it was negative. . . .

In a rehearsal of a talk on rendezvous for the recent Apollo Conference, I gave a brief reference to the plan, indicating the benefit derivable therefrom, knowing full well that the reviewing committee would ask me to withdraw any reference to this idea. As expected, this was the only item I was asked to delete.

The plan has been presented to the Space Task Group personnel several times, dating back to more than a year ago. The interest expressed has been completely negative. . . .

Grandiose plans, one-sided objections, and bias—for some inexplicable reason, everyone seems to want to avoid simple schemes. The majority always seems to be thinking in terms of grandiose plans, giving all sorts of arguments for long-range plans, etc. Why is there not more thinking in the direction of developing the simplest scheme possible? Figuratively, why not go buy a Chevrolet instead of a Cadillac? Surely a Chevrolet gets one from one place to another just as well as a Cadillac, and in many respects with marked advantages.

Concluding Remarks

It is one thing to gripe, another to offer constructive criticism. Thus, in making a few final remarks I would like to offer what I feel would be a sound integrated overall program. I think we should:

1. Get a manned rendezvous experiment going with the Mark II Mercury.[1]
2. Firm up the engine program suggested in this letter and attachment, converting the booster to these engines as soon as possible.

---

[1] NASA did so; the Mark II Mercury was renamed Gemini.

3. Establish the concept of using a C-3[2] and lunar rendezvous to accomplish the manned lunar landing as a firm program.

Naturally, in discussing matters of the type touched upon herein, one cannot make comments without having them smack somewhat against NOVA. I want to assure you, however, I'm not trying to say NOVA should not be built.[3] I'm simply trying to establish that our scheme deserves a parallel front-line position. As a matter of fact, because the lunar rendezvous approach is easier, quicker, less costly, requires less development, less new sites and facilities, it would appear more appropriate to say that this is the way to go, and that we will use NOVA as a follow on. Give us the go-ahead, and a C-3, and we will put men on the moon in very short order—and we don't need any Houston empire to do it.

In closing, Dr. Seamans, let me say that should you desire to discuss the points covered in this letter in more detail, I would welcome the opportunity to come up to Headquarters to discuss them with you.

☆   ☆   ☆

Dear Mr. Houbolt:

Thank you for your letter of November 15. In reading through your arguments and supporting material for Lunar Rendezvous, I agree that you touched upon facets of the technical approach to Manned Lunar Landing which deserve serious consideration.

I appreciate the vigorous pursuit of your ideas. It would be extremely harmful to our organization and to the country if our qualified staff were unduly limited by restrictive guidelines. In this case, however, I feel confident that we are approaching the question of Manned Lunar Landing fairly and frankly and that all views are being carefully weighed in our continuing studies.

To insure that this is indeed the case, I have sent your letter and attached material to Brainerd Holmes for his evaluation and recommendations. He will contact you directly if he requires additional information related to your ideas and concepts.

Thank you again for writing me on this matter.

---

[2] The C refers to "configuration," and the 3 implies three F-1 rocket motors. Ultimately, we developed the C-5, called Saturn V.

[3] The development of NOVA was never initiated.

☆ ☆ ☆

February 5, 1967: My initial report on the Apollo 204 fire to NASA
Administrator James E. Webb

The media were insistent that the public had a right to know the circum-
stances of the Apollo fire. The question was how to satisfy this intense pres-
sure and still permit the Review Board to conduct an investigation
methodically. James Webb obtained agreement from the President and
Congress that I would be the intermediary. After my weekly visit with the
board at Cape Canaveral, I prepared a report for Webb during my return. He
in turn presented the report to the White House and some hours later to
Congress. Congress in turn released the report to the press. In this way, the
Review Board was not constrained. If their findings ultimately differed from
mine, the mistake would be my own. The following is the first of numerous
weekly reports.

The board is taking full advantage of the extensive taped data
available as well as records made prior to the accident, the present
condition of the spacecraft, and the reports of those involved in the
test. All the physical evidence and data concerned with the test were
impounded immediately following the accident. This was to assure
that no pertinent information would be lost and that no actions
would be taken except in the full context of all the data available.

As I have stated, the preliminary review of this information has
not provided any direct indication of the origin of the fire: the preliminary analyses point to the conclusion that a clear identification of
the source of ignition or of its possible source will depend upon
detailed step-by-step examination of the entire spacecraft and its
related test support equipment.

At present, the spacecraft is still mated to the unfueled launch vehicle at the pad. However, it is being prepared for removal to our industrial
area where it will be disassembled and where experts in many technical
and scientific areas can work with the physical evidence. Prior to disas-
sembly of the damaged spacecraft, an undamaged and nearly identical
(No. 014) spacecraft will be used to establish the conditions existing
prior to the accident. The 014 spacecraft was flown from the North
American [Aviation] plant in California to Cape Kennedy on Feb. 1.

The current plans are to go through a parallel, step-by-step disas-
sembly process, first, working on the undamaged vehicle and then
repeating as closely as possible the procedure on the damaged vehicle.

In addition to analyses of recorded and physical data and equipment, the board is defining a series of investigative tasks and is assigning these to teams for execution.

At 6:31.03 P.M. E.S.T. the fire was first detected. The mission was holding at T-10 minutes. Up to this time there had been only minor difficulties with the space equipment. The purpose of the hold was to provide an opportunity to improve the communications between the spacecraft and the ground crew.

Lieutenant Colonel Grissom was the command pilot, sitting in the left seat; Lieutenant Colonel White, the senior pilot, sitting in the middle seat; and Lieutenant Commander Chaffee, the pilot, was in the right seat. . . .

At 6:31.03, pilot Chaffee reported that a fire existed in the spacecraft. . . .

At 6:31:12, or nine seconds after the first indication of fire, the cabin temperature started to increase rapidly, and pilot Chaffee reported that a bad fire existed in the cabin. Also at this time pilot Chaffee increased the illumination of the cabin lights and actuated the entry (internal) batteries. . . .

At 6:31:17, or 14 seconds after the fire was first detected, the cabin pressure reached a level of approximately 29 p.s.i. (pounds per square inch) and the cabin ruptured. . . .

The official death certificates for all three crew members list the cause of death as asphyxiation due to smoke inhalation due to the fire.

During my meetings with the board a number of other items of information were discussed, but I believe that the data I have outlined include all events having a significant bearing on an understanding of the accident.

☆　　☆　　☆

October 5 and 12, 1967: Letters from President Lyndon B. Johnson and Vice President Hubert H. Humphrey upon my resignation from NASA

After I came to the conclusion that I should resign for both NASA's good and my own, the question was how to leave on a positive note—or in Jim Webb's words when commenting on someone's departure from government amidst name-calling and rancor, how to leave without "dirtying one's nest." To arrive at the right answer, I contacted my brother Peter in Boston and former NASA counsel Walter Sohier in New York. We met on the third-floor deck of our Georgetown house. All agreed it was time for me to retire and that the letter of resignation should center on the seven years served—when only two had been planned. Gene typed the letter, which I hand-delivered to Jim Webb the next day. Two of the responses follow. Humphrey's letter refers to Edward C. Welsh, who was the executive secretary of the National Aeronautics and Space Council.

THE WHITE HOUSE

WASHINGTON

October 5, 1967

Dear Dr. Seamans:

I regret your decision to leave Government service, and I accept your resignation with a reluctance born of your own hard work and high achievement.

The responsibility and vigor with which you have carried forward our nation's space program will remain an inspiration and incentive to your colleagues and to those who will come after you. Your loyalty and devotion to the public trust has earned you the respect and gratitude of all your fellow Americans.

Please accept my appreciation for your selfless dedication to duty throughout these many years, and my very best wishes for your continued success.

Sincerely,

Lyndon B. Johnson

☆   ☆   ☆

THE VICE PRESIDENT

WASHINGTON

October 12, 1967

Dear Bob:

I was distressed to learn of your plans to leave the National Aeronautics and Space Administration, which you have served so ably. It has often been said that no one is indispensable to his government, but I assure you that some can be spared less readily than others. You are one of those who will be missed the most.

Ed Welsh told me about your calling to let me know of your decision and also passed along the information that you had planned on being with NASA just two years—and that was back in 1960. So I add patience, loyalty, and industry to the long list of competences in which you must be rated so highly.

I know that you have been a pillar of strength to Jim Webb, bringing him continuity, technical ability, managerial wisdom, and splendid personality.

Wherever you go, you will have going with you the gratitude and respect of those who know you and your work here in the government.

Sincerely,

Hubert H. Humphrey

☆   ☆   ☆

August 4, 1969: Letter from me as secretary of the Air Force to Vice President Spiro T. Agnew

In 1969 President Nixon formed a Space Task Group under the chairmanship of Vice President Spiro Agnew. The other members of the group were Thomas Paine, administrator of NASA; Lee DuBridge, the President's science advisor; and myself, recently appointed secretary of the Air Force. Mel Laird, secretary of defense, deputized me to be the Defense Department's representative to the group. At the time of our first meeting, I had been away from NASA for little more than a year; yet I found my views quite at odds with those of my former associates. Following its Apollo triumphs, NASA was actively planning Mars expeditions. My year at MIT since leaving the government had convinced me that there was little public support for such an endeavor. The views expressed in my letter were not fully accepted.

Dear Mr. Vice President:

The Department of Defense has carried out a comprehensive study of the various opportunities for using space technology to enhance national security. Options for increased space activity have been carefully reviewed by the Services, the Joint Chiefs, and the offices of the Secretary of Defense, and are the basis for a report that is being transmitted to you by Secretary Laird. As a member of your

Space Task Group, I am writing this letter to give you certain of my own personal views.

Rocketry and advanced electronics have permitted us to accomplish unique missions in this decade. The landing of the Apollo 11 astronauts on the moon and their safe return to earth is the crowning achievement. However, NASA and DoD have accomplished many other highly significant missions that are important for scientific, technical and operational reasons. . . .

We should capitalize on NASA's great scientific and technical capability to the maximum extent possible. By this I mean that NASA should wherever possible carry out work of direct relevance to man here on earth. ESSA [now NOAA] of the Department of Commerce needs assistance to understand and predict the weather more accurately for longer periods of time. The Departments of Commerce, Interior, and Agriculture need support that can be supplied by satellites if they are to carry out their responsibilities in such fields as oceanography, hydrology, agriculture, ecology, etc. However, I am not only thinking of further satellite development, but also the use of NASA's capability wherever pertinent to current national problems.

NASA should put increased emphasis in aeronautics. We, in the Department of Defense, have need for greater effort by NASA to support us in the development of military aircraft. The Department of Transportation needs major support if they are to implement a new air traffic control system. . . .

The applications of the NASA program are far reaching and considerably more effort should be expended to make the results available for the benefit of mankind. . . .

In a continuing manned lunar landing program it is important to proceed on a careful step-by-step basis reviewing scientific information from one flight before going to the next and using unmanned spacecraft where appropriate. . . .

The present Apollo Applications Program including relatively few missions should be expanded to include longer duration flights and a wider variety of orbits. . . .

I recommend that we embark on a program to study by experimental means including orbital tests the possibility of a Space Transportation System that would permit the cost per pound in orbit to be reduced by a substantial factor (ten times or more). . . .

Even though the development of a large manned space station

appears to be a logical step leading to further use and understanding of the space environment, I do not believe we should commit ourselves to the development of such a space station at this time. . . .

The unmanned planetary program should be expanded to include more thorough investigation of Mars and Venus, as well as exploration of the more distant planets.

I don't believe we should commit this Nation to a manned planetary mission, at least until the feasibility and need are more firmly established. . . .

Let us take the initiative and use the good will, the momentum and the skills demonstrated in Apollo to help solve many of our problems at home and abroad. But let us not give up exploration, rather let us also continue our exploration while validating its benefit to all mankind.

April 1 and 5, 1971: My mid-term letter to Secretary of Defense Melvin Laird and his response

While serving as secretary of the Air Force had its rewards, the sword of Damocles known as Vietnam was always hanging overhead. I became aware of it the first time I talked with Mel Laird, and it was in part the reason for my individual discussions with our four oldest children prior to my accepting the job. After these discussions, I checked one final time with Gene, who was hospitalized. Then I called Mel to tell him of my acceptance. To the best of my knowledge, there was no exchange of letters; rather, there was planning for an announcement—on January 6, 1969—of my appointment along with those of Stanley Resor, secretary of the Army, and John Chafee, secretary of the Navy.

I agreed to serve in the Nixon administration for a minimum of two years. At the end of that period, Deputy Secretary of Defense Dave Packard was planning to return to California, and Mel said he would recommend my nomination as Packard's replacement. As stated in Chapter 3, that didn't fly. Soon thereafter, our family headed for our annual outing at Vail, Colorado, and I vowed I would use the vacation to think and talk with family members about remaining in the Air Force. I can remember reading to Gene, Kathy, and Lou a longhand draft of a letter to Mel Laird regarding the continuation of my Air Force appointment, as the four of us drove to the airport in Denver. The contents were accepted not only by the family but, as can be seen from the April 5 response, by Mel Laird himself.

April 1, 1971

Dear Mel:

As you noted in a recent conversation, I agreed to join this administration for a minimum of two years, and I have no regrets in that decision. Although much remains to be done, I believe we have made significant progress phasing down our activities in Southeast Asia and improving the management of our weapons systems.

You asked me for a commitment to serve another two years. I have found this decision difficult indeed. As I indicated, there are strong family pressures influencing me to leave the government after nearly ten years in NASA and the Air Force.

If I am to stay, these pressures must be more than balanced by the goals that I can help accomplish. A Service Secretary is not in the military chain of command and hence has little to say about tactical and strategic operations. Rather a Service Secretary's job is normally limited to acquiring men and material.

Before leaving the Air Force I would like to place the C-5 contract with Lockheed on a sound basis, resolve the F-111 cost and technical difficulties, proceed with new programs such as the F-15, B-1, AWACS, A-X, and F-5E to the point where we can be reasonably confident in our policy of "fly before buy," and improve our military and civilian manpower policies. I believe another year would be required to make further significant progress.

In surveying the national mood, I am struck again and again by the concern in this country over our activities in SEA [Southeast Asia] and by the need for terminating our military involvement there as soon as possible. I believe a rapid phaseout is essential to the health and stability of this country. You have worked imaginatively toward this objective with your concept of Vietnamization. However, the plans for step by step closing of Air Force bases in South Vietnam and Thailand are still undetermined. I believe such plans and their implementation are more crucial to the country than any specific weapons acquisition or personnel projects.

My willingness to stay in DoD hinges on this administration's determination to terminate our military activities in SEA. If I continue as Secretary of the Air Force, I would want to have an opportunity to play a meaningful role in achieving this goal.

I will be happy to discuss this matter with you further.

Most sincerely,

Robert C. Seamans, Jr.

☆ ☆ ☆

THE SECRETARY OF DEFENSE
WASHINGTON, D.C.

April 5, 1971

Honorable Robert C. Seamans, Jr.
The Secretary of the Air Force

Dear Bob:

I am pleased, and even excited, to see from your April 1 letter that you are still entertaining the idea of staying with the Defense team through 1972. I recognize that you originally agreed to join the team for a two-year stint. I recognize, too, the personal and family sacrifices that are involved in such a demanding position as that of Secretary of the Air Force. It is nonetheless increasingly true that we need leaders of your rare capacity and demonstrated ability if we are to have any reasonable chance of attaining our major national goals in the Defense area.

It is only fair, of course, that you would want assurances concerning your role as Service Secretary during the coming months. There is no doubt in my mind that your role has been a central and significant one during the past two years. I have no doubt it will continue to be.

In a major sense, the Service Secretaries play the key role in determining and effecting our national security strategy. I have always felt that strategy was best defined as the aggregate of policies by which limited or finite resources were allocated to attain an established range of goals. The Service Secretaries can, and do, participate in a major way in determining those allocation policies—for all of the resources made available to Defense. The Service Secretaries can play an even larger role, even to the extent of helping determine the national goals and the aggregate resources to be made available to Defense. I welcome and solicit such a role by you, the Secretary of the Army, and the Secretary of the Navy.

You are entitled to justifiable pride in the achievements of the Air Force under your stewardship for the past two years. You have made substantial progress, inter alia, in modernization of our forces, management of key weapons systems programs, personnel planning, and important domestic action programs. Most importantly, you have played an instrumental part, perhaps greater than you realize, in moving our Southeast Asia programs in productive directions. You can be of continuing and even greater help in this vital area in the future. I share your conviction, as you know, that implementing the Nixon Doctrine in Southeast Asia is crucial. Your ability to help with the implementation of that Doctrine is unique, both from a personal and organizational standpoint. I assure you that you can and do play a meaningful role in achieving our Southeast Asia goals and, in particular, winding down US military participation in that war. In this role, I welcome and solicit an even greater participation on your part than in the past.

There are many tasks confronting the Air Force, the Defense establishment, and the nation which can be fulfilled only with your type of leadership. I am not so naive as to think that we will be able to discharge all those tasks in the next two years. To make the most substantive progress possible, however, I need you, the administration needs you, and the country needs you.

Sincerely,

Mel Laird

☆   ☆   ☆

April 10 and May 15, 1973: My letter of resignation from the Air Force and President Richard M. Nixon's response

My last six months in the Pentagon were pretty "squirrelly." As mentioned in Chapter 3, all presidential appointees were asked to resign soon after Nixon's reelection. At a meeting of the appointees in the Department of Defense, we all agreed to write one-sentence letters of resignation which the secretary of defense would hold until we truly wanted to resign. I had expected to resign earlier in the year, but when the opportunity to become president of Sloan-Kettering vanished, I decided to remain longer. As it turned out, I stayed until my appointment as president of the National Academy of Engineering.

I confess that I found the President's response to my resignation quite surprising. Nixon's chief of staff H.R. Haldemann had wanted me fired and wouldn't let me attend the special White House gala celebrating the return of the prisoners of war (POWs) from Vietnam that occurred several weeks after I joined the National Academy. Furthermore, at a luncheon for the President at the Pentagon given by Secretary of Defense Elliot Richardson several weeks before my resignation, Mr. Nixon had appeared upset, emotional, and at times irrational—quite a contrast to the thoughtful letter sent to me soon afterwards. Perhaps the preparation of the letter fell to those in the National Security Council at the White House with whom I still had reasonably good rapport.

April 10, 1973

My dear Mr. President,

It is with deep regret that I am submitting my resignation as Secretary of the Air Force to be effective in early May. As I have advised Secretary Richardson, the members of the National Academy of Engineering have asked me to be their next President. After careful consideration, I find the job is challenging and one that I must accept.

Of course, I know that no assignment could be more rewarding than serving in your Administration the past four years. As a result of your leadership, tremendous strides have been made lessening world tensions and creating improved understanding between nations. Although all issues have not been resolved in Southeast Asia, our ability to withdraw our military forces completely from South Vietnam, and the return of our POWs, is a tremendous accomplishment.

The modernization of weapon systems is another area that is vital to national security. Here again progress has been made that I'm certain will continue. The application of advanced technology has led to significant improvement in the capability of our military services and those of our allies.

Finally, let me say that it has been an honor to work with Secretary Laird, Secretary Richardson, and yourself on these matters that are so important to our country.

Yours respectfully,

Robert C. Seamans, Jr.

☆ ☆ ☆

THE WHITE HOUSE
WASHINGTON

May 15, 1973

Dear Mr. Secretary:

It is with deep regret that I accept your resignation as Secretary of the Air Force, effective on May 14, 1973, as you requested.

In doing so, I want to express my sincere appreciation for your outstanding contributions to our Nation, both as Air Force Secretary and in your previous position with the National Aeronautics and Space Administration. We have been indeed fortunate to have a man of your leadership and managerial ability directing the development of sophisticated new aircraft and helping to improve our missile systems. In an era in which technological innovation has become increasingly costly, it is to your lasting credit that the Air Force modernization programs you initiated have remained very close to target cost estimates.

Perhaps more importantly, you have recognized the importance of the individual in the Air Force and have worked to create an environment in which each member of the Air Force team feels he can realize his potential. In that effort, as well as in the other programs you have helped shape, you have laid the foundations for a stronger, more effective Air Force for many years to come.

As you return to private life, it is a pleasure to have this opportunity to express this Nation's gratitude for your distinguished service. My warmest best wishes go with Mrs. Seamans and you in your challenging new undertaking.

Sincerely,

Richard M. Nixon

☆ ☆ ☆

January 18, 1977: My letter to James Schlesinger, assistant to the President-elect (for energy)

As my time in the government ended, I sympathized with some of my predecessors because I experienced some of their frustrations. Keith Glennan prepared himself, but never had an opportunity to share his experiences with the Kennedy administration. Similarly, I had no chance to talk with Jim Schlesinger, who was destined to become the first secretary of the Department of Energy. As I indicated in my letter to him, I was prepared to stay for a short period, and a number of congressmen hoped I would, including Speaker of the House, Thomas P. ("Tip") O'Neill.

On the night of January 17, 1977, while I was attending the first show-ing of my son Joe's National Geographic Society documentary about the Hokule'a, Schlesinger called. He said he had considered asking me to stay on, but had concluded that I didn't have enough political support. I tried to con-vey some of my thoughts "without prejudice" in the letter I wrote him the next day. It was never answered.

January 18, 1977

Dr. James R. Schlesinger
Assistant to The President–Elect
School of Advanced International Studies
Johns Hopkins University
1740 Massachusetts Ave., N.W.
Washington, D.C. 20036

Dear Jim:

I have submitted my resignation to be effective January 20. Although I recognized that it was unlikely that I would be asked to remain permanently in the Carter Administration, I thought I might be of assistance for a limited period until my successor was selected. This possibility was one of the reasons that prompted me to call you last week. Incidentally, I know that a number of mutual friends have contacted you regarding my status. You should know that in no case did I ask them to be an intermediary.

I also wanted an opportunity to chat with you about the nation's energy program. It is impossible to cover many points in a short let-ter; however, in certain circles I am categorized as a nuclear advocate against conservation. I wanted you to know first hand that I have stressed the need for conservation in making decisions at ERDA and in hundreds of speeches and reports going back to "U.S. Energy Prospects: An Engineering Viewpoint" issued by the National Academy of Engineering when I was President. Conservation is, however, a difficult nettle to grasp because it impacts on and is impacted by every facet of our society. The definition, planning, and programming of conservation activities starting from scratch two years ago has been tedious, frustrating, and—at times—disappointing. Today we have a sound program and a competent dedicated staff. Their work can and should be expanded.

I am enclosing a budget sheet showing the R&D appropriation we inherited in 1975 by line item, along with the FY 76, 77, and 78 budgets for which we must take some measure of responsibility. I am also enclosing a chart from this year's planning document that shows the maximum estimated impact of energy technology in the years 1985 and 2000. Clearly, conservation is essential—but so is the expanded use of present fuels and by the year 2000 and beyond, we must rely increasingly on new fuels. The scale is quads per year which can readily be converted to millions of barrels per day simply by dividing by 2.

Although I intend to clean up my affairs and remove my junk from the office during the next week, I will take no official action after January 20th. I am delighted that you are keeping Bob Fri "without prejudice" and am certain that he will serve you well. Our line of succession goes from Bob Fri and me to the other Presidential Appointees by seniority. Hence, Jim Liverman will be the Acting Administrator from January 20 until Bob Fri returns the 23rd.

I expect to be in Washington at least for the next few months. If you care to discuss any energy issues, I will accommodate to your schedule.

Good luck and best wishes,

Sincerely,

Robert C. Seamans, Jr.
Administrator

☆   ☆   ☆

January 18 and 19, 1977: My letter of resignation from ERDA to President Gerald R. Ford, his reply, and a January 19 letter to me from Richard W. Roberts of ERDA

My two years as administrator of ERDA were the most hectic and jam-packed of my life. It started in May 1974 when I was contacted by the White House personnel office to discuss the possibility of serving as administrator of an as-yet nonexistent agency. After a series of meetings extending over several months, I met with Frank Zarb, then in the Office of Management and Budget, later to become administrator of the Federal Energy Administration (FEA). He was convinced that the enabling legislation for ERDA would be passed by Congress and wanted me to commit to be its administrator. I told him my commitment would have to await a conversation with the President. The President turned out to be Ford, not Nixon.

I received a call asking me to appear for President Ford's announcement of my nomination on Halloween morning. When I explained that I had not yet met with the President, I had the meeting that afternoon, and the announcement was made the following day, November 1, 1974. There was no time for an exchange of letters. My meeting with President Ford is described in Chapter 4. My departure from ERDA was equally abrupt, starting with my letter of January 18, 1977, and President Ford's response the following day.

I have also included a letter from Dick Roberts (dated the same day as Ford's letter), whom I "shanghaied" from the National Bureau of Standards and who directed the most controversial ERDA department (nuclear energy).

January 18, 1977

The President
The White House

Dear Mr. President:

It has been my privilege to serve in your Administration as the Administrator of the Energy Research and Development Administration.

As you have often stated, the United States must conserve energy and develop alternate fuels in order to become less dependent on imported oil—an expensive and wasting resource. A major national effort is required that galvanizes government, industry, and the public into concerted, dedicated action. You have stressed the need to resolve the anticipated technical, social, economic, and political issues, and have forwarded to the Congress many sound recommendations including a plan for a new Department of Energy.

Although much remains to be done in the field of energy, this is the logical time for a change in the leadership of ERDA. We have developed a strong organization with competent, professional personnel and many challenging programs. Hence it is with sadness, but a measure of satisfaction that I submit my resignation to be effective January 20, 1977.

Respectfully yours,

Robert C. Seamans, Jr.
Administrator

☆ ☆ ☆

THE WHITE HOUSE
WASHINGTON

January 19, 1977

Dear Bob:

Thank you for your letter of January 18. I, of course, accept your resignation as Administrator of the Energy Research and Development Administration, effective January 20, 1977, as you requested. In doing so, I would like to take this opportunity to commend you for the dedication and commitment you have brought to our efforts to achieve our goal of energy independence. Tremendous challenges lie ahead, but with your help and that of others in my Administration I am confident that we have laid a solid foundation for completing the enormous task which confronts our Nation.

I am indeed grateful for the contributions you have made and for the professional manner in which you have carried out your many and varied responsibilities. You will always be able to look back with great pride on your Government service and on the fine record you have compiled.

As you prepare to return to private life, you may be sure that you take with you my best wishes for every success and happiness in your future endeavors.

Sincerely,

Gerald R. Ford

☆   ☆   ☆

January 19, 1977

Dr. Robert C. Seamans, Jr.
Administrator

Dear Bob:

I'll never forget the day when I came down to 7th & D to sell you on the National Bureau of Standards' energy conservation programs, only to find myself sold on ERDA's nuclear programs. You gave me a great personal and professional opportunity and for this I thank you.

But of even more value to me than the importance and challenge of the job was the opportunity to work with and for you. Leadership really can't be written about—it can only be observed and experienced. In a very short period you took a disparate collection of entrenched people and programs, acquired an unusually talented group of people from many sources, established the confidence of the Congress and business community, and in a quiet yet forceful way created ERDA.

You once told me your major job was to knock down barriers so we could get the job done. This you have done well.

I have learned a great deal from you, Bob, on how to accomplish things in this infinitely complex system, such as:

- analyze, and really know where you want to go,
- be flexible and resilient,
- be cool and calm with Congress,
- be prepared (super prepared!),
- be humble,
- be tenacious,
- have a thick skin (very thick!),
- smile—it helps,
- etc. . . .

Your influence on ERDA and the national energy program has been profound. You should be very proud of ERDA's accomplishments and the organization which will remain to carry out its work.

Best wishes for success, good health and happiness! ERDA, the Country, and I, personally, will sincerely miss you.

Sincerely,

Richard W. Roberts
Assistant Administrator for Nuclear Energy

☆ ☆ ☆

December 16, 1985: Letter to me from former President Gerald R. Ford upon his resignation from the Aerospace Corporation

Gerald Ford was such a good guy, and the experience of having a former President as a trustee during my chairmanship of the Aerospace board was so unique, that I am including this letter, written nearly nine years after the end of my government service. Aerospace Corporation is nonprofit and contracts exclusively with the Department of Defense. The corporation provides engineering and planning services for all national security activities in space. Jerry was a trustee for three years, retiring at the mandatory age of seventy-two.

At his first trustees' meeting, he arrived in the boardroom after the other trustees were in place. I walked around with him, introducing each member to "Jerry Ford." Some complained I was too informal, but he did not.

On another occasion, we were having a retirement ceremony for a key employee. Seeing a handsome, apparently new briefcase near the podium, I presented it to the retiree. It wasn't until after the departure of the retiree following the ceremony that Jerry indicated that he would like his briefcase returned. This we did, buying a similar one for the retiree.

GERALD R. FORD

December 16, 1985

Dear Bob:

I am deeply grateful for the opportunity to have served on the Board of Trustees of Aerospace Corporation under your leadership. It was an interesting and most enjoyable experience. Thank you for your many kindnesses.

It is my hope that our paths will cross again soon. As you know I have the highest admiration for your dedicated public service over many years in positions of highest responsibility in our federal government. Most of all, I am indebted to you for your superb service in my White House Administration.

On a personal basis, I deeply appreciate your friendship which I treasure.

Warmest, best wishes for a wonderful Holiday Season for you and your family.

Jerry Ford

Biographical Appendix

Richard Borda (1931–) was assistant secretary of the Air Force for Reserve Affairs, 1970–1973.

Harold Brown (1927–) was director of Defense Research and Engineering at the Pentagon, 1961–1965, before becoming secretary of the Air Force, 1965–1969. After spending eight years as the president of the California Institute of Technology, he returned to Washington to serve as the secretary of defense in the Carter administration, 1977–1981. He currently works at the Center for Strategic and International Studies in Washington.

John Chafee (1922–) was the secretary of the Navy, 1969–1972. In 1976, he was elected to the U.S. Senate as a Democrat from Rhode Island and has served there since.

Roger Chaffee (1935–1967) was a Navy lieutenant commander and astronaut who had never flown in space. Chaffee, along with his crewmates Gus Grissom and Ed White, were killed when their Apollo 204 capsule was engulfed in flames on the launch pad at the Kennedy Space Center.

Leighton (Lee) Davis was an Air Force lieutenant general who served as the National Range Division commander from 1960 to 1967. He received a Distinguished Service Medal for his role as the Department of Defense manager for the Mercury and Gemini programs.

Kurt Debus (1908–1983) was a German engineer who came to the United States in 1945 with a group of engineers and scientists headed by Wernher von Braun. After working at Fort Bliss, Texas, and the Redstone Arsenal in Huntsville, Alabama, Debus moved to Cape Canaveral, Florida, where he supervised the launching of the first ballistic missile fired from there. Debus became director of the Launch Operations Center and then of the Kennedy Space Center, as it was renamed in December 1963. He retired from that position in 1974.

Charles Stark (Doc) Draper (1901–1987) earned his Ph.D. in physics at the Massachusetts Institute of Technology in 1938 and became a full professor there the following year. In that same year, he founded the Instrumentation Laboratory. Its first major achievement was the Mark 14 gyroscopic gunsight for Navy anti-aircraft guns. Draper and the lab applied gyroscopic principles to the development of inertial guidance systems for airplanes, missiles, submarines, ships, satellites, and space vehicles—notably those used in the Apollo Moon landings. Draper was a mentor to many future students in aerospace engineering.

Hugh Dryden (1898–1965) was a career civil servant and an aerodynamicist by discipline who had begun life as something of a child prodigy. He graduated at age 14 from high school and earned an A.B. in three years from Johns Hopkins (1916). Three years later (1919), he earned his Ph.D. in physics and mathematics from the same institution, even though he had been employed full time by the National Bureau of Standards since June 1918. His career at the Bureau of Standards, which lasted until 1947, included becoming its assistant director and then associate director during his final two years there. Dryden served as director of the NACA, 1947–1958, after which he became deputy administrator of NASA under T. Keith Glennan and James E. Webb.

Maxime Faget (1921–), an aeronautical engineer with a B.S. from Louisiana State University (1943), joined the staff at Langley Aeronautical Laboratory in 1946 and soon became head of the Performance Aerodynamics Branch of the Pilotless Aircraft Research Division. In 1958 he joined the Space Task Group in NASA, forerunner of the NASA Manned Spacecraft Center (later renamed the Johnson Space Center), became its assistant director for engineering and development in 1962, and later its director. He contributed many of the original design concepts for Project Mercury's crewed spacecraft and played a major role in designing virtually every U.S. crewed spacecraft since that time, including the Space Shuttle. He retired from NASA in 1981 and became an executive for Eagle Engineering, Inc. In 1982 he was one of the founders of Space Industries, Inc., and became its president and chief executive officer.

Robert Gilruth (1913–) was a longtime NACA engineer who worked at the Langley Aeronautical Laboratory, 1937–1946, then as chief of the Pilotless Aircraft Research Division at Wallops Island, 1946–1952. He had been exploring the possibility of human spaceflight before the creation of NASA. He served as assistant director at Langley, 1952–1959, and as assistant director (crewed satellites) and head of Project Mercury, 1959–1961—technically assigned to the Goddard Space Flight Center but physically located at Langley. In early 1961, T. Keith Glennan established an independent Space Task Group (already the group's name as an independent subdivision of Goddard) under Gilruth at Langley to supervise the Mercury program. This group moved to the Manned Spacecraft Center in Houston in 1962. Gilruth was then director of the Houston operation, 1962–1972.

John Glenn (1921–) earned a B.S. in engineering from Muskingum College and became a colonel in the Marine Corps. A member of NASA's first class of astronauts, Glenn was the first American to orbit the Earth, which he did in 1962 on the Mercury–Atlas 6 (*Friendship* 7) mission. First elected to the U.S. Senate in 1975, he is still a Democratic Senator from Ohio.

T. Keith Glennan (1905–1995) served as the first administrator of the National Aeronautics and Space Administration from August 1958 to January 1961. Glennan had worked in the sound motion picture industry in the 1930s and joined the Columbia University Division of War Research in 1942. In 1947 he became president of the Case Institute of Technology. From October 1950 to November 1952, he served as a member of the Atomic Energy Commission. Upon leaving NASA in 1961, Glennan returned to Case, where he continued to serve as president until 1966.

Harry Goett (1910–) was an aeronautical engineer who began working at the Langley Aeronautical Laboratory in 1936 and then worked at the Ames Aeronautical Laboratory, 1948–1959. In 1959 he became director of the Goddard Space Flight Center, a post he held until July 1965, when he became a special assistant to NASA Administrator James E. Webb. Later that year, he shifted over to the private sector, working at Philco's Western Development Laboratories in California and then at Ford Aerospace and Communications.

Virgil (Gus) Grissom (1927–1967) was chosen for the first group of astronauts in 1959. He was the pilot for the 1961 Mercury-Redstone 4 (*Liberty Bell 7*) mission, a suborbital flight; the command pilot for Gemini III; and the backup command pilot for Gemini VI. He had been selected as commander of the first Apollo flight at the time of his death in the Apollo 204 fire in January 1967.

Grant Hansen (1921–) was the assistant secretary of the Air Force for Research and Development, 1969–1973.

D. Brainerd Holmes (1921–) was involved in the management of high technology efforts in private industry and the federal government. He was on the staff of Bell Telephone Laboratories, 1945–1953, and RCA, 1953–1961. He then became deputy associate administrator for manned space flight at NASA, 1961-1963. Holmes left NASA to work for the Raytheon Corporation.

John Houbolt (1919–) was an engineer who worked as an aircraft structures specialist at NASA's Langley Research Center. After President Kennedy announced his 1961 decision to put an American on the Moon, Houbolt was instrumental in the technical decision to adopt the lunar-orbit rendezvous approach for the Apollo program. Houbolt left NASA in 1963 for the private sector, but he returned to Langley in 1976 before retiring in 1985.

Hubert Humphrey (1911–1978) (D–MN) served as a U.S. senator from Minnesota, 1949–1964 and 1971–1978. As senator, he pressed for the creation of a cabinet-level Department of Science and Technology in early 1958, which was defeated by President Eisenhower's proposal to establish NASA. He was Vice President of the United States, 1965–1969, under Lyndon Johnson, but he lost the presidential election to Nixon in 1968.

John Johnson (1915–) served as general counsel of the Air Force, 1952–1958. He accepted the same position at NASA in 1958. In 1963 he left NASA to join the Communications Satellite Corporation. He retired in 1980.

Lyndon Johnson (1908–1973) (D–TX) was a U.S. senator, 1949–1960, Vice President of the United States, 1960–1963, and President, 1963–1969. Best known for the social legislation he passed during his presidency and for his escalation of the war in Vietnam, he was also highly instrumental in revising and passing the legislation that created NASA and in supporting the U.S. space program as chair of the Committee on Aeronautical and Space Sciences and of the preparedness subcommittee of the Senate Armed Services Committee. He later chaired the National Aeronautics and Space Council (as Vice President under President Kennedy).

David C. Jones (1921–) joined the Air Force during World War II and advanced through the ranks, becoming the deputy commander of operations in Vietnam, the Air Force chief of staff, 1974–1978, and finally the chair of the Joint Chiefs of Staff, 1978–1982.

John Kennedy (1916–1963) was President of the United States, 1961–1963. In 1960, as a senator from Massachusetts (1953–1960), he ran for President as the Democratic candidate, with Lyndon Johnson as his running mate. On May 25, 1961, President Kennedy announced to the nation the goal of sending an American to the Moon before the end of the decade. The human spaceflight imperative was a direct outgrowth of it; Projects Mercury (at least in its latter stages), Gemini, and Apollo were each designed to execute it.

Robert Kerr (1896–1963) (D–OK) was governor of Oklahoma, 1943–1947, and then was elected to the Senate the following year. From 1961 to 1963, he chaired the Committee on Aeronautical and Space Sciences.

James Killian, Jr. (1904–1988), who was president of the Massachusetts Institute of Technology (MIT), 1949–1959, took leave between November 1957 and July 1959 to serve as the first presidential science advisor. President Dwight D. Eisenhower established the President's Science Advisory Committee, which Killian chaired, following the Sputnik crisis. After leaving the White House staff in 1959, Killian continued his work at MIT, but in 1965 he began working with the Corporation for Public Broadcasting to develop public television.

Melvin Laird (1922–) was secretary of defense, 1969–1973, during the Nixon administration. He later served on the boards of directors of a number of major corporations.

George Low (1926–1984) was an Austrian aeronautical engineer who joined the NACA in 1949. He became chief of manned spaceflight at NASA Headquarters in 1958. In 1960 he chaired a special committee that formulated the original plans for the Apollo lunar landings. In 1964 he became deputy director of the Manned Spacecraft Center in Houston, the forerunner of the Johnson Space Center. He became deputy administrator of NASA in 1969 and served as acting administrator, 1970–1971. He retired from NASA in 1976 to become president of the Rensselaer Polytechnic Institute, a position he held until his death. In 1990 NASA renamed its quality and excellence award after him.

John McLucas (1920–) was the under secretary of the Air Force, 1969–1973, and then secretary, 1973–1975. From 1975 to 1977, he served as the administrator of the Federal Aviation Administration.

Robert McNamara (1916–) was secretary of defense during the Kennedy and Johnson administrations, 1961–1968. Thereafter, he served as president of the World Bank, where he remained until retirement in 1981. As secretary of defense in 1961, McNamara was intimately involved in the Kennedy administration's process of approving Project Apollo.

Walter Mondale (1928–) was Vice President of the United States under President Jimmy Carter (1977–1981). He ran for President himself in 1984 but lost to incumbent Ronald Reagan. Mondale served in the Senate as a Democrat from Minnesota, 1964–1977, and was considered a harsh critic of large technology programs such as the Space Shuttle. He currently serves as the Clinton administration's ambassador to Japan.

George Mueller (1918–) was associate administrator for manned spaceflight at NASA Headquarters, 1963–1969, where he was responsible for overseeing the completion of Project Apollo and beginning the development of the Space Shuttle. He left NASA for private industry in 1969.

David Packard (1912–1996) was a co-founder of the Hewlett-Packard Company. He was deputy secretary of defense in the Nixon administration, 1969–1971.

Thomas Paine (1921–1992) was appointed deputy administrator of NASA in 1968, acting administrator later that year, and then NASA's third administrator in 1969. During his leadership, the first seven Apollo manned missions were flown. Paine resigned from NASA in September 1970 to return to the General Electric Company, where he remained until 1976. In 1985 the White House chose Paine as chair of a National Commission on Space to prepare a report on the future of space exploration. The Paine Commission report, *Pioneering the Space Frontier*, was released in May 1986. It espoused a "pioneering mission for 21st-century America"—"to lead the exploration and development of the space frontier, advancing science, technology, and enterprise, and building institutions and systems that make accessible vast new resources and support human settlements beyond Earth orbit, from the highlands of the Moon to the plains of Mars." The report also contained a "Declaration for Space" that included a rationale for exploring and settling the solar system and outlined a long-range space program for the United States.

Samuel Phillips (1921–1990) was an electrical engineer who had a distinguished military flying record during World War II. He became involved in the development of the successful B-52 bomber in the early 1950s and headed the Minuteman intercontinental ballistic missile program in the latter part of the decade. In 1964 Phillips, by this time an Air Force general, moved to NASA to head the Apollo lunar landing program, which, of course, was unique in its technological accomplishment. He returned to the Air Force in the 1970s and commanded the Air Force Systems Command prior to his retirement in 1975.

William Proxmire (1915–) (D–WI) served as a U.S. senator from Wisconsin, 1957–1989. He was well known for his congressional investigations of government waste and abuse of funding.

Stanley Resor (1917–) was the secretary of the Army during the Johnson and Nixon administrations, 1965–1971.

John (Jack) Ryan was an Air Force general who became the Air Force chief of staff.

Spencer Schedler was assistant secretary of the Air Force for financial management.

Julian Scheer (1926–) was a newspaper reporter who served as NASA's assistant administrator for public affairs, 1962–1971.

Willis Shapley (1917–), son of famous Harvard astronomer Harlow Shapley, joined the Bureau of the Budget in 1942 and held increasingly more responsible positions in military and space affairs at that agency for more than 20 years. In 1965 he moved to NASA as associate deputy administrator, with his duties including supervision of the public affairs, congressional affairs, DOD and interagency affairs, and international affairs offices. He retired in 1975 but rejoined NASA in 1987 to help it recover from the *Challenger* disaster. He served as associate deputy administrator (policy) until 1988, when he again retired but continued to serve as a consultant to the administrator.

Joseph Shea (1926–) joined NASA Headquarters' Office of Manned Space Flight in 1962. The next year, he was named the Apollo program manager at the Manned Spacecraft Center in Houston. In 1967 he returned to NASA Headquarters as deputy associate administrator for manned spaceflight. He joined the Raytheon Company in 1968 and served on the NASA Advisory Council for several years. Shea returned to NASA again as head of Space Station redesign efforts in the early 1990s, and he also served as chair of a task force that reviewed plans for the first servicing mission of the Hubble Space Telescope.

Alan Shepard (1923–) was the first U.S. astronaut in space. Following cosmonaut Yuri Gagarin's first spaceflight in April 1961, Shepard flew on a short suborbital flight in May 1961. He also flew to the Moon on Apollo 14 in 1971. He is currently the president of Seven Fourteen Enterprises in Houston.

Abe Silverstein (1908–) worked as an engineer at the Langley Aeronautical Laboratory, 1929–1943, and moved to the Lewis Laboratory (later Research Center) to a succession of management positions, the last (1961–1970) as director of the center. When T. Keith Glennan arrived and NASA began in 1958, Silverstein was director of the Office of Space Flight Development. While at Headquarters, he helped create and direct the efforts leading to the spaceflights of Project Mercury and to establish the technical basis for the Apollo program. As director of Lewis, he oversaw a major expansion of the center and the development of the Centaur launch vehicle. He retired from NASA in 1970 to take a position with the Republic Steel Corporation.

Jack Stempler (1920–) was the assistant secretary of defense for legislative affairs, 1965–1970 and 1977–1981. In between (1970–1977), he served as the general counsel for the Air Force.

Curtis Tarr (1924–) was the assistant secretary of the Air Force for manpower and reserve affairs, 1969–1970. He then became the director of the Selective Service System, 1970–1972.

Albert Thomas (1898–1966) (D–TX) chaired the House Independent Offices Appropriations subcommittee that had jurisdiction over NASA. First elected to Congress in 1936, he ran this powerful subcommittee for almost 15 years. He used his political influence to have the $250 million Manned Spacecraft Center located in Houston, near his congressional district.

Floyd Thompson (1898–1976) joined the Langley Aeronautical Laboratory in 1926 as part of a staff of only about 150. He became chief of the Flight Research Division in 1940 and assistant chief of research for Langley in 1943. In 1960 Thompson became director of Langley. He also served briefly as a special assistant to the NASA administrator in 1968 before retiring later that year.

Wernher von Braun (1912–1977) was the leader of the "rocket team" that had developed the German V-2 ballistic missile in World War II. At the conclusion of the war, von Braun and some of his chief assistants came to Fort Bliss in El Paso, Texas, to work on rocket development and use the V-2 for high-altitude research. In 1950 von Braun's team moved to the Army's Redstone Arsenal near Huntsville, Alabama. From 1960 to 1970, he was the director of NASA's Marshall Space Flight Center, where he was instrumental in supervising the Saturn rocket program for the Apollo lunar missions.

James Webb (1906–1992) was NASA administrator between 1961 and 1968. Previously, he had been an aide to a congressman and a business executive with the Sperry Corporation and the Kerr-McGee Oil Company. He also had been director of the Bureau of the Budget, 1946–1950, and undersecretary of state, (1950–1952).

Edward Welsh (1909–) had a long career in various private and public enterprises. He served as legislative assistant to Senator Stuart Symington (D–MO), 1953–1961, and was the executive secretary of the National Aeronautics and Space Council through the 1960s.

Edward White (1930–1967) was the first astronaut to "walk" in space, which he did in 1965 as part of the Gemini IV mission. A lieutenant colonel in the Air Force and son of an Air Force general, White joined NASA in 1962 as a member of its second class of astronauts. White was killed, along with crewmates Roger Chaffee and Gus Grissom, when their Apollo 204 capsule was engulfed in flames on the launch pad at the Kennedy Space Center.

Philip Whittaker was NASA's assistant administrator for industry affairs in the 1960s. President Nixon later appointed him assistant secretary of the Air Force for installations and logistics.

Jerome Wiesner (1915–1994) was science advisor to President John F. Kennedy. He had been a faculty member of the Massachusetts Institute of Technology and had served on President Eisenhower's Science Advisory Committee. During the presidential campaign of 1960, Wiesner had advised Kennedy on science and technology issues and prepared a transition team report on the subject that questioned the value of human spaceflight. As Kennedy's science advisor, he tussled with NASA over the lunar landing commitment and the method of conducting it.

Source: Biographical reference files, NASA Headquarters History Office, Washington, D.C.

Chronology for Robert C. Seamans, Jr.

October 30, 1918	Born in Salem, Massachusetts
1940	Received bachelor's degree from Harvard College
1941–1955	Held various teaching and research positions at the Massachusetts Institute of Technology (MIT)
1942	Received master's degree from MIT
June 13, 1942	Married Eugenia (Gene) Merrill
1951	Received doctor of science degree from MIT
1955	Started work at RCA as manager of its Airborne Systems Laboratory
1958	Chief engineer at RCA's Missile Electronics and Controls Division
1960	Named NASA associate administrator
1965	Named NASA deputy administrator
1968	Returned to MIT as professor in aeronautics and astronautics
1969	Appointed secretary of the Air Force
1973	Named president of the National Academy of Engineering

1974   Appointed first administrator of the Energy Research
                and Development Administration

1977   Returned to MIT as Henry Luce professor of environ-
                ment and public policy

1978   Became dean of MIT School of Engineering

1981   Resigned as dean to pursue private sector activities

1984   Retired as Henry Luce professor of environment and
                public policy at MIT

1984–1996   Senior lecturer in aeronautics and astronautics at MIT

INDEX

THE NASA HISTORY SERIES

Reference Works, NASA SP-4000

Grimwood, James M. *Project Mercury: A Chronology*
(NASA SP-4001, 1963)

Grimwood, James M., and Hacker, Barton C., with Vorzimmer, Peter
J. *Project Gemini Technology and Operations: A Chronology*
(NASA SP-4002, 1969)

Link, Mae Mills. *Space Medicine in Project Mercury*
(NASA SP-4003, 1965)

*Astronautics and Aeronautics, 1963: Chronology of Science,
Technology, and Policy* (NASA SP-4004, 1964)

*Astronautics and Aeronautics, 1964: Chronology of Science,
Technology, and Policy* (NASA SP-4005, 1965)

*Astronautics and Aeronautics, 1965: Chronology of Science,
Technology, and Policy* (NASA SP-4006, 1966)

*Astronautics and Aeronautics, 1966: Chronology of Science,
Technology, and Policy* (NASA SP-4007, 1967)

*Astronautics and Aeronautics, 1967: Chronology of Science,
Technology, and Policy* (NASA SP-4008, 1968)

Ertel, Ivan D., and Morse, Mary Louise. *The Apollo Spacecraft: A
Chronology, Volume I, Through November 7, 1962*
(NASA SP-4009, 1969)

Morse, Mary Louise, and Bays, Jean Kernahan. *The Apollo
Spacecraft: A Chronology, Volume II, November 8,
1962–September 30, 1964* (NASA SP-4009, 1973)

Brooks, Courtney G., and Ertel, Ivan D. *The Apollo Spacecraft: A
Chronology, Volume III, October 1, 1964–January 20, 1966*
(NASA SP-4009, 1973)

Ertel, Ivan D., and Newkirk, Roland W., with Brooks, Courtney G.
*The Apollo Spacecraft: A Chronology, Volume IV, January 21,
1966–July 13, 1974* (NASA SP-4009, 1978)

*Astronautics and Aeronautics, 1968: Chronology of Science,
Technology, and Policy* (NASA SP-4010, 1969)

Newkirk, Roland W., and Ertel, Ivan D., with Brooks, Courtney G.
Skylab: A Chronology (NASA SP-4011, 1977)

Van Nimmen, Jane, and Bruno, Leonard C., with Rosholt, Robert L. *NASA Historical Data Book, Vol. I: NASA Resources, 1958–1968* (NASA SP-4012, 1976, rep. ed. 1988)

Ezell, Linda Neuman. *NASA Historical Data Book, Vol II: Programs and Projects, 1958–1968* (NASA SP-4012, 1988)

Ezell, Linda Neuman. *NASA Historical Data Book, Vol. III: Programs and Projects, 1969–1978* (NASA SP-4012, 1988)

Astronautics and Aeronautics, 1969: Chronology of Science, Technology, and Policy (NASA SP-4014, 1970)

Astronautics and Aeronautics, 1970: Chronology of Science, Technology, and Policy (NASA SP-4015, 1972)

Astronautics and Aeronautics, 1971: Chronology of Science, Technology, and Policy (NASA SP-4016, 1972)

Astronautics and Aeronautics, 1972: Chronology of Science, Technology, and Policy (NASA SP-4017, 1974)

Astronautics and Aeronautics, 1973: Chronology of Science, Technology, and Policy (NASA SP-4018, 1975)

Astronautics and Aeronautics, 1974: Chronology of Science, Technology, and Policy (NASA SP-4019, 1977)

Astronautics and Aeronautics, 1975: Chronology of Science, Technology, and Policy (NASA SP-4020, 1979)

Astronautics and Aeronautics, 1976: Chronology of Science, Technology, and Policy (NASA SP-4021, 1984)

Astronautics and Aeronautics, 1977: Chronology of Science, Technology, and Policy (NASA SP-4022, 1986)

Astronautics and Aeronautics, 1978: Chronology of Science, Technology, and Policy (NASA SP-4023, 1986)

Astronautics and Aeronautics, 1979–1984: Chronology of Science, Technology, and Policy (NASA SP-4024, 1988)

Astronautics and Aeronautics, 1985: Chronology of Science, Technology, and Policy (NASA SP-4025, 1990)

Gawdiak, Ihor Y. Compiler. *NASA Historical Data Book, Vol. IV: NASA Resources, 1969–1978* (NASA SP-4012, 1994)

Noordung, Hermann. *The Problem of Space Travel: The Rocket Motor.* In Ernst Stuhlinger and J.D. Hunley, with Jennifer Garland, editors (NASA SP-4026, 1995)

Management Histories, NASA SP-4100

Rosholt, Robert L. *An Administrative History of NASA, 1958-1963* (NASA SP-4101, 1966)

Levine, Arnold S. *Managing NASA in the Apollo Era* (NASA SP-4102, 1982)

Roland, Alex. *Model Research: The National Advisory Committee for Aeronautics, 1915-1958* (NASA SP-4103, 1985)

Fries, Sylvia D. *NASA Engineers and the Age of Apollo* (NASA SP-4104, 1992)

Glennan, T. Keith. *The Birth of NASA: The Diary of T. Keith Glennan,* edited by J.D. Hunley (NASA SP-4105, 1993)

Seamans, Robert C., Jr. *Aiming at Targets: The Autobiography of Robert C. Seamans, Jr.* (NASA SP-4106, 1996)

Project Histories, NASA SP-4200

Swenson, Loyd S., Jr., Grimwood, James M., and Alexander, Charles C. *This New Ocean: A History of Project Mercury* (NASA SP-4201, 1966)

Green, Constance McL., and Lomask, Milton. *Vanguard: A History* (NASA SP-4202, 1970; rep. ed. Smithsonian Institution Press, 1971)

Hacker, Barton C., and Grimwood, James M. *On Shoulders of Titans: A History of Project Gemini* (NASA SP-4203, 1977)

Benson, Charles D. and Faherty, William Barnaby. *Moonport: A History of Apollo Launch Facilities and Operations* (NASA SP-4204, 1978)

Brooks, Courtney G., Grimwood, James M., and Swenson, Loyd S., Jr. *Chariots for Apollo: A History of Manned Lunar Spacecraft* (NASA SP-4205, 1979)

Bilstein, Roger E. *Stages to Saturn: A Technological History of the Apollo/Saturn Launch Vehicles* (NASA SP-4206, 1980)

Compton, W. David, and Benson, Charles D. *Living and Working in Space: A History of Skylab* (NASA SP-4208, 1983)

Ezell, Edward Clinton, and Ezell, Linda Neuman. *The Partnership: A History of the Apollo-Soyuz Test Project* (NASA SP-4209, 1978)

Hall, R. Cargill. *Lunar Impact: A History of Project Ranger* (NASA SP-4210, 1977)

Newell, Homer E. *Beyond the Atmosphere: Early Years of Space Science* (NASA SP-4211, 1980)

Ezell, Edward Clinton, and Ezell, Linda Neuman. *On Mars: Exploration of the Red Planet, 1958-1978* (NASA SP-4212, 1984)

Pitts, John A. *The Human Factor: Biomedicine in the Manned Space Program to 1980* (NASA SP-4213, 1985)

Compton, W. David. *Where No Man Has Gone Before: A History of Apollo Lunar Exploration Missions* (NASA SP-4214, 1989)

Naugle, John E. *First Among Equals: The Selection of NASA Space Science Experiments* (NASA SP-4215, 1991)

Wallace, Lane E. *Airborne Trailblazer: Two Decades with NASA Langley's Boeing 737 Flying Laboratory* (NASA SP-4216, 1994)

Butrica, Andrews J. *To See the Unseen: A History of Planetary Radar Astronomy* (NASA SP-4218, 1996)

Center Histories, NASA SP-4300

Rosenthal, Alfred. *Venture into Space: Early Years of Goddard Space Flight Center* (NASA SP-4301, 1985)

Hartman, Edwin, P. *Adventures in Research: A History of Ames Research Center, 1940-1965* (NASA SP-4302, 1970)

Hallion, Richard P. *On the Frontier: Flight Research at Dryden, 1946–1981* (NASA SP-4303, 1984)

Muenger, Elizabeth A. *Searching the Horizon: A History of Ames Research Center, 1940-1976* (NASA SP-4304, 1985)

Hansen, James R. *Engineer in Charge: A History of the Langley Aeronautical Laboratory, 1917-1958* (NASA SP-4305, 1987)

Dawson, Virginia P. *Engines and Innovation: Lewis Laboratory and American Propulsion Technology* (NASA SP-4306, 1991)

Dethloff, Henry C. *"Suddenly Tomorrow Came . . .": A History of the Johnson Space Center, 1957–1990* (NASA SP-4307, 1993)

Hansen, James R. *Spaceflight Revolution: NASA Langley Research Center From Sputnik to Apollo* (NASA SP-4308, 1995)

General Histories, NASA SP-4400

Corliss, William R. *NASA Sounding Rockets, 1958–1968: A Historical Summary* (NASA SP-4401, 1971)

Wells, Helen T., Whiteley, Susan H., and Karegeannes, Carrie. *Origins of NASA Names* (NASA SP-4402, 1976)

Anderson, Frank W., Jr. *Orders of Magnitude: A History of NACA and NASA, 1915–1980* (NASA SP-4403, 1981)

Sloop, John L. *Liquid Hydrogen as a Propulsion Fuel, 1945–1959* (NASA SP-4404, 1978)

Roland, Alex. *A Spacefaring People: Perspectives on Early Spaceflight* (NASA SP-4405, 1985)

Bilstein, Roger E. *Orders of Magnitude: A History of the NACA and NASA, 1915–1990* (NASA SP-4406, 1989)

Logsdon, John M., with Lear, Linda J., Warren-Findley, Jannelle, Williamson, Ray A., and Day, Dwayne A., eds. *Exploring the Unknown: Selected Documents in the History of the U.S. Civil Space Program, Volume I: Organizing for Exploration* (NASA SP-4407, 1995)

Logsdon, John M., with Day, Dwayne A., and Launius, Roger D., eds. *Exploring the Unknown: Selected Documents in the History of the U.S. Civil Space Program, Volume II: External Relationships* (NASA SP-4407, 1996)